"As we reached for the stars during Apollo, we stood on the shoulders of giants—of Americans like David Cisco who believed not only in themselves but in God and their country. None of our accomplishments would have been possible without the dedication, effort, self-sacrifice and commitment of these men and women from all cultures of American society.

"David knows that what you believe in, you can make happen—what you dream can become a reality. In his book, *Full Circle,* he shows he was willing to seek the knowledge, make the commitment and take the risk."

—Captain Gene Cernan, USN (ret.);
commander, Apollo XVII;
Last man on the moon

"What an inspiring story. I think it should be required reading in high school, especially for those who might pursue technical careers. Maybe get NASA to put it on their reading list...especially the astronauts. It also makes the older reader look inward and has real lessons about how hard the folks behind the scenes work. I really enjoyed the read."

—Richard Allen, Jr., president and CEO,
Space Center Houston

"I thoroughly enjoyed reading David's book, *Full Circle,* which provides an engaging look into one of the people who took us to the moon more than forty years ago. In a very entertaining and readable fashion, David has included so many important lessons for leading a fulfilling and successful life. Everyone should take the opportunity to read this book and learn from it!"

—Bill Foster, lead shuttle (GC) flight controller,
Johnson Space Center

FULL CIRCLE

An incredible journey
of a Lunar Module Spacecraft Technician,
his memoirs of his time at NASA
and all the stories along the way

David L. Cisco
Apollo Alumni
Lunar Module Spacecraft Technician

DLC Enterprises

Book design by:
Arbor Books, Inc.
www.arborbooks.com

Printed in the United States of America

Full Circle
David L. Cisco

1. Title 2. Author 3. Memoir

Library of Congress Control Number: 2009914371

ISBN 10: 0615345638
ISBN 13: 978-0-615-34563-5

This book is dedicated to my mother,
Elnora Fredana Cunningham, and mother-in-law,
Mabel Suddin—along with my wife, Amina; son, Kumar;
daughter, Lahna; my grandchildren, Xavier, Anika, and Hayes.

ACKNOWLEDGMENTS

Thanks a bunch to Kathianne Boniello, the publishing company, and all the people behind the scenes—and, of course, the person who always believed in me and set my compass in the right direction, Tom Short.

PART I

Maryanne!

Thanks for visiting

NASA

Donna Icasia :)

I was startled by the sudden, loud knock on my front door. I couldn't fathom who on earth would be bothering me—or anyone—on this special day.

I rarely got visitors to my small, Houston-area apartment. After all, I was at work more often than not, and it was only recently that my fiancée had joined me in my adopted city. It was July 20, 1969, late on a Sunday afternoon, and we, like hundreds of millions of people around the world, had been glued to the television.

I was a little apprehensive as I tried to think of who could be out in the hall.

Pulling myself away from the history being beamed into the living room, I walked to the door. But when I peeked outside, I was a bit surprised to see an older, slightly frail man. He was balding and unobtrusive-looking. Dressed in a neat uniform, he immediately announced, in a slightly firm voice:

"Western Union."

He handed me an envelope.

Now I was definitely nervous. None of my friends, or anyone on my old block in New York, would be delivering me a telegram. In fact, I had never even gotten a telegram before.

The deliveryman could probably tell that I was perplexed, though I was doing my best to act like this was no big deal.

I thanked him, slowly signing his clipboard before shutting the door. I was in my early twenties at the time, and it never even occurred to me that I should have tipped the man.

By now, my heart was in my stomach and my chest felt like a heard of elephants was stampeding across it. There's no other way to describe the anticipation. After all, didn't telegrams usually indicate bad news? What could it possibly be?

I tore quickly into the envelope, trying desperately to get to the body of the telegram before it finally came free. I pulled out a slightly yellow paper, featuring just a few lines written in red:

David,

Congratulations on sending a man to the moon.
I'm so proud of you.

Love,
Mom

I was flooded with relief. Thank God no one had died! But it wasn't too long before the feeling of gratitude was overtaken by something else. I was inside my apartment but, at that moment, I was a little embarrassed. I couldn't help but wonder as I stared at my mother's expression of pride: *Just how much bragging had I done?*

I hoped my mother didn't think that this was something I, or anyone, could have done single-handedly—I hoped I hadn't been giving that impression! But it wasn't long before I chuckled to myself, and realized that this was just my mom being a mom, doing what every parent has a right to do and taking a little joy in the accomplishments of their children.

In truth, I could hardly claim real credit. I didn't go to space myself, nor did I fly in any rockets or man mission control. Like almost everyone else, I was home watching Neil Armstrong and Buzz Aldrin walk across the surface of the moon.

But it wasn't just a novelty for me, as I watched that

fuzzy television feed. I felt an intimate connection with those astronauts and their journey to outer space aboard the lunar module, what we called the LM for short.

I hadn't planned on being a part of one of the most exciting, innovative and important chapters in American history. But then again, the nearly 400,000 men and women who worked in what came to be known as the Apollo program would probably say the same thing.

During the Apollo program, which dominated the 1960s and 1970s, there were approximately thirty-two astronauts and ten flight directors, whose names and faces everyone came to know.

And then there were those hundreds of thousands of everyday workers, who came from all ethnic, economic and educational backgrounds, to help answer President John F. Kennedy's challenge to the nation on May 25, 1961. It was on that day that Kennedy told the United States Congress:

> …I believe that this nation should commit itself to achieving the goal, before this decade is out, of landing a man on the Moon and returning him safely to the Earth. No single space project in this period will be more impressive to mankind, or more important in the long-range exploration of space; and none will be so difficult or expensive to accomplish.

The Apollo program helped bring our country's hopes and dreams to new heights during a chaotic time in our culture and politics. It was a series of missions and journeys to space brought on by a large workforce striving to bring scientific and mechanical advancements to life.

In doing so, the men and women of the Apollo program,

those who were famous and those who were not, worked together to create a history of which we can all be proud.

That's why this story should resonate with thousands of people who don't stand out from the crowd, don't lead organizations and don't play starring roles but who nevertheless strive to be a part of something big.

Maybe my mother was right—because without those 400,000 everyday workers, of whom I was one, the moon landing would never have been possible. Behind the famous astronauts and flight directors, there was practically an army of people like me—people who spent countless hours flipping switches and turning dials, running tests, checking equipment and laying the groundwork for the missions.

And who was I? I didn't come from wealth or have a fancy education. I had no special connections to open doors for me within NASA.

I was just a kid from Long Island, New York, who probably had more obstacles than opportunities while growing up. If someone was trying to predict what would become of me, I don't think they would have chosen me to be a part of history. But that didn't stop me from being a part of one of the country's most magnificent, magical accomplishments. I wasn't in the limelight, but I helped send people to the stars. And in doing so, I found my own way to shine.

But we didn't think about stardom, in any form, back when I was growing up. We didn't think that far ahead. I grew up in a quaint little community called Oyster Bay, about an hour outside of New York City—but it wasn't a childhood of green lawns, picket fences and suburban comfort.

We think of those days, the 1950s and early 1960s, as simpler times, and in many ways they were, but that didn't mean that life's complexities and burdens were lessened. We just handled them differently.

In many ways my youth was about the simple things. At times, there simply wasn't enough to go around. Not enough space; not enough time for my parents to spend with myself and my five siblings; not enough coal for the fire; not enough money; sometimes, not enough food.

Those are some of the facts of my youth, but not necessarily the truth of it.

The truth was, even with eight people to one bathroom and two sets of bunk beds crammed into a back bedroom for my brothers and me, our house never felt cramped.

Even though the home, along with a few others, was somewhat isolated in a semi-industrial area of town, we never felt lonely or unsafe.

And though some might have considered us as being from the wrong side of the tracks, if you will, we never lived that way or felt that way, either.

Our small rented house was at 9 Harbor Place, which paints a pretty picture if you think about it, but, in reality, we lived on an unpaved road, unofficially dubbed "the alley."

The alley was a dead-end street, a dirt road with potholes and a single streetlight. Next door was a small factory, as well as a few neighbors nearby, who, like us, were renting their houses, and behind our home was a junkyard.

To the right of our home was Theodore Roosevelt Memorial Park, which was sometimes visited by bums—though the bums of yesteryear were generally of a more polite variety than most of the people forced to live on the streets today.

Adjacent to the alley was a piece of property used by the Long Island Rail Road, the area's major commuter train system. The Long Island Rail Road shuttled people from all over Long Island to and from work in New York City, just as it does today. The property near our house was the end of the line for the Oyster Bay branch of the LIRR, and that was the place where they used to turn the trains around for the return trip to the city.

It was pretty convenient for my father, Floyd Townsend Cisco, who worked as a car inspector for the LIRR and took the train to and from work every day.

Tall and thin, my father was a sharp dresser who served in World War II as part of the U.S. Army. He was born on October 27, 1918, and was, as we described it back then, a colored man, with Shinnecock Indian ancestry in his blood. For years he was the picture of health and always worked and provided for his family: his wife, Elnora Fredana Constance Cunningham, and their six children, which included myself, four brothers and a sister.

My mother was of Irish descent, with a quick wit and the ability to freeze each and every one of her children with a single, all-powerful, all-knowing look. Two years younger than my father, my tiny mother—all ninety pounds of her—was without a doubt the disciplinarian, the one you couldn't slip anything past.

She kept an immaculate home, and we were always busy doing chores, cleaning our rooms, working in some way or another. It was never questioned; it was just expected of us and we did what we were told.

The house wasn't exactly elaborate: a small stoop and enclosed porch led into the kitchen, which had a small pantry and wood-burning stove. There was a dining room and, just off to the right, a living room, with a few bedrooms upstairs. Four of us boys shared one of those small rooms, two sets of bunk beds doing the job. My sister, Elaine, the second oldest and only girl, who delighted in mothering us boys, had her own room. I never could figure that one out.

Elaine was, of course, our mother's pride and joy, and often helped take care of her younger brothers, making sure we were washed up and clean before dinner. My oldest brother, Floyd—well, we all knew he was the favorite. He always seemed to be given a free pass. I can't ever remember him being in the doghouse for anything.

The mischief-maker was generally my brother Ronnie. I always remember my mother saying: "Ronnie, sit down," or "You shouldn't have done that," or "What are you doing?"

Ronnie was only a year older than me. My brother Wayne and I were mostly the kids stuck in the middle, while the baby of the family was my brother Marvin, who we called Skip.

One of my brothers was inevitably left out of the upstairs bunk bed equation, and for years we'd shift furniture around in the dining room at the end of the night, and he would sleep there. The one thing that was out of the question was the living room—that was formal, kept for company. My mother made sure we never dared step foot inside.

The funny thing was we rarely had visitors. Perhaps that's because my parents were trying to protect us.

My parents gave me, gave all of us, another central truth. We were an interracial family during a time when, factually speaking, that situation might have drawn controversy or ridicule. But I honestly never felt racial tension at or near our home. Our parents sheltered us from that, managed to keep such ugliness out.

By the time I started going to school, the other kids made sure I knew that I was different in some ways. My olive skin tone and "good" hair, as we used to say, drew questions I wasn't sure how to answer, questions that it had never occurred to me to ask.

"What are you?"

"Where is your mother from? Where is your father from?"

I would always just answer that I was American, because, after all, I was. I never thought I looked any different, but I guess I did. They were questions that resurfaced over the years that never really seemed to go away.

I never thought I—or any of us—was different, and never felt that way. And though my mom had her hands full caring for our family, there were no excuses for, or ignoring of, bad behavior.

I found that out the hard way.

This is a secret I've carried with me almost my whole life, but for many years I was on probation.

My offense? Shoplifting.

My judge, jury and probation officer? My mother—a more frightening court of law than anything the American judicial system ever produced.

I don't want to incriminate anyone in the retelling of my crime, but it was a day that started innocently enough. I was seven years old, and one of my brothers and I—I'll call him R.C., just to be safe—were walking to the center of town. We were going to the five and dime.

At the time, marbles were a huge part of my world. I was a pretty good marble player, and I had a particular affection for those marbles that looked like they had certain shapes inside, like a cat's eye.

We walked into the store and came to a huge display of marbles packaged behind glass. We stood ogling the precious, gemlike pieces of glass, looking at the different colors and types. It didn't take me long to realize that some of the marbles—four, to be exact—were loose.

I asked R.C., "What are those?"

"Oh, they're just extras," he said, moving along.

Extras? Well if they're just extras, that meant nobody really needed them, right? I figured that if they were extras, it would be okay if I used them. After all, they were cat's eyes, my favorite kind.

So I quickly scooped up the "extras" and put them in my pocket, satisfied with my find and running to catch up with R.C.

We went about our business and walked on home. It was when we got there that the trouble began.

With six kids, my mom had a lot to keep track of, but she always kept a sharp eye on all of us. She always knew what we had and where we had gotten it.

She spotted my "extra" marbles in about the space of a single heartbeat, and zeroed in on me like a scud missile zooming in on a target. It took about five seconds before I broke, confessing how I had acquired my new treasures.

Now, I was a young kid but I wasn't stupid. And if it had been my father handling the situation, I'd often found that a little logic would sometimes work to smooth things over and get me out of a tight spot.

I truly felt I hadn't done anything wrong. They were extras! They probably never would have been sold anyway, and I could have, and intended to, put them to good use.

But that didn't fly with my mom.

She was only five-foot-two, but my mom was a strong woman. She moved like lightening with a force of emotion to match her physical speed, grabbing my hand and snatching the marbles away before I knew what the heck was happening.

Her anger was clear as she plopped my marbles into her empty purse and immediately marched me out of the house and down our little stoop.

We walked right back to the five and dime and, during the half-mile trip, I could hear my precious cat's eye marbles clinking and knocking together in her bag.

Entering the store, my mother wasted no time in asking for the manager. When he came out, he was wearing a tie with a big knot in it, similar to what Kmart managers wear today.

There we were: me with these adults standing over me. My mother made me confess my sin again and give up my stash. I reluctantly handed over my cat's eyes, apologizing and promising never to do it again.

The manager was a nice man; he seemed like maybe he even felt a little sorry for me. He was willing to let me off the hook.

I was relieved and glad that part of the ordeal was over. It occurred to me that I wasn't sure what might happen when we got back home, but I thought I'd deal with that when we got there.

My mom had other ideas.

Before I knew it, she had stuck her head out of the store, calling the local police officer off his beat.

As he walked over, I couldn't quite believe it—again, I had to repeat what I had done, this time making my confession to the officer, sealing it with yet another apology, in front of everybody.

At that point, my mother's mission was accomplished. I got her message. Believe me, I got it!

We walked home after that, and I was very much under my mom's watch. She kept a close eye on everything I did, and made it clear that what I had done would never be okay.

That was my first and last offense of that nature. My probation officer—my mom—made sure of that, and I made sure never to let her down again.

Another kid might have taken that experience an entirely different way. I never felt like my mother was being overbearing or extreme in her response to my marble fiasco. I never resented my mom or became angry with her for what she did. Instead, I knew I was the one who was out of line, and as time went on I felt a unique closeness with my mother, a trust forged out of what became a tight bond between the two of us.

No matter the times, no matter the situation, family can find a way to bring us back to ourselves, to keep us on the path to where we need to be. There were so many reasons I could have failed, so many diversions and distractions that could have turned me and my siblings down a dark path.

But things didn't turn out that way. Part of it was the innocence of the times, perhaps; part of it was simply who we were; a lot of it was my parents.

We didn't live a luxurious life, and while we may have secretly wished for things in our hearts, we never whined or complained. It just wasn't done. It took years before we, as children, fully comprehended our family's situation.

Of course, we didn't have today's technology—the cell phones, the portable DVD players, the fancy cars or video games.

We used everything we had, everything that was available to us.

I remember, at the age of ten, grabbing a bucket and going out with my brothers to hunt the grounds near the railroad yard. The locomotives were powered by large chunks of black coal that the train personnel stored behind the engine compartment. Sometimes, they'd overfill them and coal would spill out and fall onto the ground.

Coal was something we would usually pay for, so we took the railroad's waste for ourselves. You know what they say about one man's trash being another's treasure. My brothers and I would fill our buckets and bring the coal back home, and it would be used for heating the house.

Collecting coal was one of the ways our family tried to make ends meet. We had to, not just because we were a large group, but because going to war had left my father a changed man.

He was proud to serve his country, but the war brought

a fundamental shift in his health that had devastating consequences. I was not born until 1945, so I remember only what I was told.

When my father was honorably discharged from the Army and returned to the United States after the war, he was almost immediately beset by migraines.

The blistering pain of a migraine headache could last for days at a time. The nausea and sensitivity to light, sound and motion, along with a throbbing pain, could be crippling. Back then there wasn't much anybody could do, or even knew, about the condition. It was a bit of a mystery, and not widely understood. The migraines proved to be disabling for my father, and he suffered with them for years.

If there wasn't much sympathy from the general population, most of whom probably never had direct experience with migraines, the Veterans Administration didn't help much, either. The VA hospitals were supposed to care for ailing military veterans, but they weren't exactly receptive to blacks.

A hardworking family man, my father was virtually marooned by himself on an island of pain and confusion with little to no help available. The migraines eventually affected his balance and depth perception. That meant that sometimes he appeared to be walking funny—the damage left by the migraines would make a curb seem closer than it was, for example.

Today, they at least have medicines that can prevent the awful episodes or help migraine sufferers, but my father had no such tools on which to rely.

He often had no choice but to seek out a dark, quiet corner to ride out the intense pain, and in the summer had no way to escape the stifling heat that must have only compounded an already unimaginable ache.

It was so unfair, so unjust. And the migraines created situations that my father shouldn't have had to endure.

My parents were unique in many ways, not the least of which was the fact that I never once saw them drink or smoke. It wasn't that they were particularly religious—we were Baptist, but were never really regular churchgoers. Alcohol just never touched either of my parent's lips, and I never even heard my father so much as say a curse word or anything disrespectful about anybody. I tried to follow that example in my life by respecting everyone's point of view, and you will never hear me use profanity, which is my small way of honoring my father. Cursing just wasn't in my father's nature; it wasn't who he was.

One day he was coming home on the train when he was overcome by a migraine. The motion and noise of the train, and knowing there was no way to escape it, could only have made the horrible pain worse.

Skip had gone to stand on the platform to greet our dad.

As the train doors slid open and he made his way off, you could see my father had to have been in the grips of a severe migraine. The pain in his head was practically incapacitating him, and he had trouble walking off the train.

Skip took his hand and tried to walk with him, help him make it off the platform and to the house. We lived so close but every step must have been agony, must have felt like a mile.

It was then that a policeman approached Skip and my father.

The officer began barking out commands almost immediately. He removed Skip's tiny hand from my father's hand, and moved the small boy away. Then he roughly slammed my father to the ground.

If it was possible to feel terror through the blinding haze of pain my father was surely in, I am sure my father was consumed by it. I cannot imagine what was going through his mind as he lay face first on the ground.

As the police officer handcuffed my father, he clearly never stopped to take a closer look at the situation. Maybe he just saw a black man having trouble walking and assumed that what he had encountered was a stumbling, bumbling drunk who needed to be thrown in a cell for a few hours.

I don't know how long my father was lying on the ground, his hands yanked and locked behind him by the officer's handcuffs. At that age, I'm sure I couldn't have comprehended what might have happened had another train not pulled up to the platform.

But another train did come in and, thankfully, there was a man on that train who knew my father from work. The man saw my dad lying there and had the good sense to quickly speak up.

"That's Floyd," he said. "He doesn't drink."

I'm sure the police officer didn't understand at first. I'm sure he must have done a double-take. Maybe it was because the man on the train was white, or maybe his statements were enough to get the officer to stop and think, but he eventually, reluctantly, took the handcuffs off and allowed my father to get up.

My father managed to scrape himself off the platform. He collected my frightened brother, or maybe Skip got him, and the two made their way home. But the incident was a haunting reminder of the times, of the racial dynamic my parents tried so hard from which to shield us. And I believe that incident is at least one reason my father never ventured out too much or too far from home afterwards.

Usually, the migraines meant he couldn't work. His bosses wouldn't allow him to return without a medical clearance, something the doctors wouldn't issue.

The immense physical toll of the migraines and the inability to work and support his family like he used to, before he had gone into the Army, had a profound effect on my father.

As the years progressed I could see how bad he felt, how much his inability to provide for his family undermined who he was. It was sad to see how, much of the time, it seemed as if he were a beaten down man, someone who was just getting through each day as best as he could without making trouble and accepting, as much as he could, the way things were.

My mother worked when she was able to, usually cleaning houses. But the harsh reality was that after a while, without a steady, significant income, we were forced to go on public assistance.

I'm sure it galled my father to have to go on welfare. I'm sure it was never what my father and mother dreamed about or planned for their children. What parent would?

I can remember, once a month, there they would be: a block of yellow cheese and a package of powdered milk.

White water poured over cornflakes tastes terrible. It's no proper substitute for the real thing, and there was no way to pretend that it was. I always felt hungry during those times.

I never asked my brothers or sister how they felt, if they had the same experience or feelings I did, but it was never enough for me. However, I never said anything. After all, you only get your share and there wasn't much of anything anyone could have done about it. There was no use in whining or crying over it.

Besides, it didn't matter how it tasted, or how I felt about it. This was a safety net, cast out to support our family in a very volatile time. Without it, I wonder where we would have ended up and what would have happened to us, and I'm grateful for the protection we were afforded.

The thing was that my father was rather quiet about his troubles. I remember him always standing with his sandwich and a cup of Lipton tea; I remember being able to sometimes talk my way out of trouble if he was the one in charge. But I don't recall him complaining or sharing what must have been a bottomless well of confusion, loneliness, perhaps even resentment.

It was an example of what kind of man he was that whenever he felt well enough, my father volunteered for the civil defense in Oyster Bay. I remember him getting dressed up and going to those meetings, feeling very proud. He even earned an award from Governor Nelson Rockefeller for his five years of service with the group.

Perhaps that was the greatest gift our parents gave us: a generally positive outlook on life. They focused on what had to be done, taking every day as it came and always moving forward.

I'm sure they had doubts. I'm sure there were times they were afraid, but they didn't spend endless hours wallowing in self-pity or letting their fear impact their kids. They just kept going on, working hard and realizing that even though so many things were out of their control, they still had choices.

They chose life, one as whole and complete as possible for all of us, and in making that decision, they chose success.

Maybe, because they had so many children, they didn't feel they had a choice. That's the kind of people they were—good people.

That's the lesson they gave me, perhaps the greatest gift they could impart. The gift of knowing my own power, and knowing that no matter what happened to me, I could choose what to do and whether or not I wanted to succeed. It didn't matter what was going on around me, what obstacles might be in the way: a lack of money, moving to a strange city and being separated from those you love, endless hours of work, racism.

The power was mine to use or ignore. My fate was truly in my own hands, and it was a lesson I took in early.

I wanted to earn a little extra cash, so, as a kid, I went to work. My first job alone was something that kids probably haven't done in years. I was determined to make some extra money, so I decided to become a paperboy, and delivered Long Island's daily paper, *Newsday*, around the neighborhood.

It gave me a taste of having a few dollars in my pocket and a sense of responsibility. After all, rain or shine, the paper had to be delivered.

I wasn't really concerned about answering to the paper man who gave me the job—my real boss, after all, was my mother, the one whose expectations I really had to measure up to.

I liked having a little money of my own. I dropped the paper off each day to about thirty customers and, after a week of collecting from them, I would tally up what I owed the man at the paper and the rest, mostly the tips, would be mine.

But in my neighborhood it took a while just to track down and collect on the papers that were owed. That meant I often came up short in what I owed to my boss, and that the venture wasn't as profitable as I had hoped.

One of the places I delivered papers was the junkyard behind our house. I don't know if it really was a junkyard the way we mean that term today, but it was a place where people brought rags, newspapers, scrap metal and other items. The guys at the junkyard would weigh what people brought in and pay money for it based on the weight.

Sometimes they would sell some items on Saturdays at the local farmers' market, which I would occasionally attend. I kind of got acquainted with the junkyard guys, and they used to kid around with me.

So when it came time for me to step up the career ladder

and abandon the paper route, I had my sights set on the junkyard.

I was thirteen years old, and walked in one day looking for summer work. The guys at the junkyard wasted no time in hiring me. I worked ten hours a day, six days a week, earning twenty-five dollars a week. I don't know what the employment laws were for children at the time, but maybe they weren't applicable in Oyster Bay back then!

But it was a job that made a lot of sense at the time. I would open the gate to our yard and walk less than twenty or thirty feet, wait for the guys to open up and then to go to work. As teenage jobs go, I could have done a lot worse.

People would drive up, randomly bringing their newspaper, recycling rags or metal, and the men would use their weights and scales to decide how much they would pay out. The papers would go into a press that was in the ground, and it would create big bales of paper that would be bundled.

One of my main responsibilities was to drive the forklift. I would use the forklift to move around all the stuff, the bundles of paper, the scrap metal, etc., that people brought in and place it where it needed to be in the junkyard.

I look back now and it makes me shake my head to think of a young teen behind the wheel of a piece of machinery like that. We regard and raise our children so differently now than parents did in the 1950s and 1960s. Maybe their way was better. It's hard to say.

As people get older they give in to things like fear, low self-esteem and other people's expectations and opinions.

When you're a kid, you are uninhibited in the best way possible—you think you can do anything and if you're given the chance. You figure out that you can do almost anything.

I thought I had a pretty cool job. I would always give my

mother fifteen dollars every week to help toward the family expenses. The other ten I would mostly save, sticking the bills in a jar until I could open my own savings account.

I worked at the junkyard for a while before I actually did the math on what I was really getting paid and came to a rather discouraging result, realizing that my pay amounted to less than fifty cents an hour.

So it took just about two summers before my ambitions started to grow, and I began to look for another job.

Oyster Bay's premiere hardware store was in the center of town. Nobman's Hardware Emporium was a family-owned business that had been in Oyster Bay since 1910. It was, and still is, a local institution.

I wasn't intimidated, though. A confident teen, I remember walking in one day with my head held high and approaching the owner, Mr. Nobman. I asked if they needed any help at the store.

Well, Mr. Nobman must have had a soft spot in his heart, seeing this skinny kid coming into his shop and asking for a job. I mean, what did I know about the hardware business?

He thought about it for a few minutes. Then he looked at me and said I could start work the next day, and that I would be trained when I reported to work.

I could barely contain my happiness and thanked him profusely. I was ecstatic and feeling over the moon as I raced back home. No more junkyard! Finally, I had a respectable job.

It wasn't until I was trying to explain this next step up the career ladder to my mother that I realized one tiny little problem—and that was that I couldn't give her any details about what my new job entailed. I wasn't bothered too much, though. I only knew I would be making much more money than the minimal income the junkyard job had offered.

The next day I showed up at my brand new job, eager and excited. It was then that I was handed a large broom, which had what looked like a three-foot duster on one end. Then Mr. Nobman handed me a bucket with about a half a gallon of what appeared to be reddish saw dust in it. My responsibilities included dispensing a bit of the sawdust on the floor of the store, and then working it around until I eventually removed it.

I learned this was to hold the dust down in the store, and that my main task was to sweep the floors. It wasn't exactly glamorous work; it certainly didn't have the "cool" factor that running a forklift did.

But Mr. Nobman had a great deal of faith in me, and he always encouraged me to move forward and try something new. He never held me back or kept me just as the kid who swept the floors.

This was a real hardware store, after all, with rows and rows of tiny drawers with every conceivable nut and screw in them. The store was filled with tools and gadgets and equipment, and the customers expected you to know your stuff.

It wasn't long before I was doing more than just sweeping the floors. I was allowed to work with customers as well. The big trick was making sure you learned and remembered where everything was, especially in those small drawers, so you didn't end up opening half a dozen of them in front of the customers and appear as if you didn't know what you were doing.

Nobman's was more than just a summer job. But I never lost that can-do attitude that had helped me get so far to begin with. I just learned things from the beginning, including how to drive a stick shift. I didn't explain that I didn't know how to drive that kind of pickup truck, which the store used for deliveries. I just watched and then, when it was time, jumped into the truck, started it up and went. I figured if I started out really slowly, the boss wouldn't realize that I didn't know what I was doing!

But as long as I had the basics, I always felt that I could work out whatever problem came my way. I may have tried to start off in second gear a couple of times, but everything worked out. If I was given the opportunity, I could usually make the most of it.

I worked there for several years, before and after I entered high school. It took a while but as time went on I became quite proficient in repairing lamps, picture frames, lawnmowers, etc.

Finally, I had reached the pinnacle—I was allowed to run the cash register. I felt like I had finally arrived, to know that the owner trusted me to handle the cash. It was a big responsibility and I took it seriously.

It was around this time that international events were bringing outer space much more into our everyday world, and even though I didn't know it, shaping my future.

As I was cutting my teeth and expanding my skills working at my hometown hardware store, the Russians were sending cosmonaut Yuri Gagarin to the stars. It was April 1961, and Gagarin became the first person to fly in space.

Just a few weeks later, America made its own strides in what became the race to get into space. On May 5, 1961, astronaut Alan Shepard became the first American in space.

I was a junior high school student, and remember huddling around a black and white television set with my classmates, watching coverage of the countdown and Shepard's takeoff into space aboard his Mercury-Redstone rocket.

I don't recall how or why our class was watching it. I suppose one of my teachers made it happen. It was quite exciting to watch, even though the rabbit ear antennas on the old television and the horizontal and vertical lines on the screen made it a bit hard to see. Still, we got enough of a picture and could make out the launch and Shepard's historic flight, which sent him into a fifteen-minute sub-orbital flight. He didn't go to the moon, but it was just as exciting as if he had.

It was about this time that the international competition between America and the Soviet Union was heating up. It was partly because of the fear that the United States would be beaten to space by the Russians that President Kennedy was making his historic declaration to Congress, urging America to venture into one of the final frontiers and send a man to the moon.

It was exhilarating and overwhelming to watch that history being made, but it didn't dominate my life. I was still in high school and the future was a long way off. Or at least, that's how it felt.

And that's how it was when we went to high school. No one really talked to me about the future, or what I should—or could—do.

But it was about this time that I remember what was surely one of the worst moments of my mother's life. My mother had been the only child of Cora Mayhew. My grandmother was five-feet-six-inches tall, thin, with flaming red hair, and she made quite an impressive sight. But though she was only in her fifties, she was suffering from a life-threatening disease. For years she lived next door to us in Oyster Bay but I felt like I never really got a chance to get to know her.

She had taken a turn for the worse, and my mother had gone to the hospital in Glen Cove to visit her often. But I'll never forget the day the news came: October 14, 1960. The phone had rung and, in those days, it was a loud, howling ring like a fire department bell that reverberated through the whole house.

When she got the call, I was on the steps, just sitting there, listening. I could hear part of the conversation, and then there was silence. Then there was just sheer devastation and it was obvious my mother's heart had been broken. I'll never forget the agony and pain my mother bore when she got that call.

It was a painful time, but my grandmother had lived long enough to see her daughter grow and become the matriarch of her own family. It was part of a life that my mother had lived most of in Oyster Bay, from her school days through the growth of her own kids.

The school was in a large building that housed all the grades under one roof. My parents had gone to the same high school that my siblings and I did.

My parents had no concept of college. It was so far outside their reality. Neither one of my parents graduated from high school themselves. As they struggled to make ends meet and keep our family afloat, I understood clearly that their goal was to see that we all graduated from high school. I don't know if they ever thought it possible that we may continue our education beyond that. But in retrospect it didn't seem as if anyone in our high school was thinking that way, either. At least, when it came to kids like me.

I wasn't the only one. My longtime friend Joe Cucci had the same experience.

I had known Joe almost all of my life; we started school at the same time. He was one of my best friends, a small, wiry kid like me. Though everyone knew I lived on the other side of the tracks, Joe lived nearby, on Orchard Street. He had lost his mother when he was just thirteen years old; his father drove a truck and other heavy equipment for a living, and Joe came from a Sicilian family of hard workers. We spent a lot of time together, and maybe that explains my love of Italian food.

Joe, at times, seemed like the kind of guy who was a bit

soft, a person who would go along to get along but he could always hold his own.

Even though I had plenty of brothers already, Joe and I spent so much time together that he and I were practically family. In fact, we sealed the bond with a childhood ritual, pricking our fingers and then rubbing the small cuts together—the theory behind it being that we would now and forever be blood brothers, with the same blood running through each other's veins.

Neither of us was a standout in school, either in looks or athletics—although Joe probably had a different perspective on the subject of looks. Despite that, I was inspired to take up an extracurricular activity one Memorial Day. I was watching the Memorial Day parade coming down the main street of Oyster Bay, and the marching band was passing by me.

The unmistakable feeling imparted by a loud, enthusiastic marching band is infectious. The heart-throbbing beat of the drums caught my attention, and I started to think it would be so cool to one day be in the marching band. As they made their way in front of me, I noticed all the musicians and their instruments, which had place cards holding up the sheet music they were playing.

Except the drummers, that is. The drummers in the marching band didn't have any sheet music. It looked like they were just following the rhythm and didn't need to read it, and I figured that meant they had it easier than the rest of the band. I wasn't a jock or anything, so I decided that I would take the easy way out and become a drummer. That way, I reasoned, I wouldn't have to learn to read music.

Boy, was I wrong. At first, becoming a part of the percussion section was boring, because I just practiced with a wooden block that was covered in rubber. As it turns out, drummers have to do more than be able to read sheet music.

In the percussion section, you had to memorize the music, and I spent a lot of time doing just that, along with counting bars, because I had to be able to know exactly when to come in to the song and to do so right on cue.

It was like anything else one has to learn—a lot of hard work. But I got to the point where I could read music and play not only in the marching band, but in the orchestra as well.

Along the way I earned five medals at the New York State School Music Association, which was quite an accomplishment.

The marching band was my favorite of all the different musical groups, and I was so proud of myself. I'll never forget our music teacher, Donald Luckinbill—the kind of educator who took the time to encourage you. I only wish there had been more teachers like him.

I never understood why our school counselors didn't do a better job of seeking out kids like me and the many, many others who maybe didn't have the economic or educational opportunities that other, more well-off families had.

Someone could have taken a few extra minutes, stepped up and said: "I know your family may not have the resources to send you to college, but there are other options."

But they didn't. We were pushed to take menial courses, and to fill up the rest of the day with shop or woodwork. Surely, things weren't as competitive back then when it came to getting into a college, but, still, every child needs encouragement. If you wait until the senior year of high school and then suddenly start to overhear other students talking about where they're going to college—well, even if it dawns on you that you missed something, by then it's too late.

It seemed no one was looking out for kids who showed a willingness to learn, who showed some potential, or who maybe just needed a little help. It wouldn't have taken much.

Maybe some tutoring or simple guidance would have been enough to keep countless people from falling through the cracks of uncertainty.

That was a danger we always faced, even though we always managed to catch up. But it's harder to do so from behind.

Both Joe and I did what we had to do to get out of high school. It wasn't easy.

By that time, my family's finances were in such disarray that we were finally evicted from our home in the alley. We had lived in our small rental house for the last ten years. It was a confusing, emotional time. In order to finish school, I went from living as a part of a family of eight to being virtually independent, staying with my father's stepmother so I could complete my senior year of high school in Oyster Bay. My family then moved to the nearby town of Mineola to live in another rented house.

I kept working at Nobman's, and would take the Long Island Rail Road on weekends to see my family, but, at seventeen, I had no car of my own. We walked a lot more back then in general, so walking several miles to go to school and back didn't feel like too much of a big deal. Having a car would have made things a lot easier, but we did what we had to do without complaining. In that regard, nothing had changed in my life. The situation was what it was, and there was no use in whining about it.

But the drastic change raised a lot of questions and fears. Where would we go? What would we do? It was tough. Though I never lost contact with my "probation officer" and talked to my mother regularly, it wasn't the same as living with my family each day. I would take the train to Mineola on the weekends, and then walk the two miles to my family's new place. But I spent a lot of time alone, working until 6 p.m. on Saturdays and then rushing to catch a 7 p.m. train. And then by the time I got there, there was homework to do before getting back in time for school and work on Monday.

I was getting ready to enter my senior year of high school, and I didn't want to change schools before I had a chance to graduate. Living with my step-grandmother was a different experience. She was a kind woman and well intentioned, but it was definitely an adjustment to live by yourself with no parents or siblings. She was also a Jehovah's Witness and I remember having to sit at the table, read the books and go along with the program, but it was something I was doing out of respect and not to make waves. I never got too much heat for it; it was just part of living in her house.

I was working hard, making it through the days, but was still mostly invisible at school. I always felt like a loner. I had been working at some pretty serious jobs since I was thirteen years old and I knew the value of work, but there were times I would be working at Nobman's in that last year of high school and I got a glimpse of the other kids' lives that made me wistful.

I would have to head straight down to Main Street from school right after classes so I could get to the hardware store on time. Of course, I didn't have a car then so I had to walk the mile and a half.

Meanwhile the girls would be walking with their books in their hands, and the guys would be driving down Main Street, some in their convertibles and others in their beautiful Crown Victorias.

They were all heading to Snouders Drug Store, which was in many ways the center of town.

Snouders was famous for many years since, for a long while, it was the only place in town with a phone. When President Theodore Roosevelt stayed at his home in nearby Sagamore

Hill, the president would receive phone calls at Snouders, and the store would send someone to his estate on Sagamore Hill to get him. So Snouders was a mainstay, a focal point of the community.

And that included the kids after school, too. They would gather over at Snouders for ice cream sodas and hang out. I used to see them there because it was right across the street from Nobman's.

As soon as I got to work, one of the first things I had to do was raise or lower the canopy outside the storefront to keep the hot sun from coming into the store. I had to use a twist-type of tool on a long metal pole, and it went quite slowly. That gave me time to look out of the corner of my eye and see the kids coming down the street on the one side and, with the other, spying the people at Snouders, some of whom were watching me.

It was embarrassing. And there were times I thought, *Wouldn't it be nice to just go over there and hang out, and have an ice cream soda with my friends?* However, I didn't exactly have the money for that. It wasn't something that I focused on, but it was inescapable.

At some point I was noticed by someone at school and given a bit of encouragement, but little did I know it was the wrong message. I remember a guidance counselor telling me, "You're not college material," before pointing out that my family wouldn't be able to afford to send me to college anyway. I should just try to finish high school, he said, and then get a factory job. I've never forgotten that school counselor, and can recall his name and face as if it were yesterday.

With the country erupting around us, that wasn't what any kid needed to hear from a person who was supposed to be helping us. The 1960s were among the most turbulent decades in our nation's history and it was a frightening time to be thrown into the "real world."

President Kennedy had been assassinated in Dallas in 1963. The activist Malcolm X was killed; racial violence and riots were erupting around the country as African-Americans fought for equal rights against decades of oppression. The Vietnam conflict was unfolding, taking countless young men off to fight in a bloody, horrifying war that barely resembled the World Wars our parents had been in.

It was a time of shifting and loosening cultural values. Rock and roll was becoming more prominent, as was drug use. The upheaval of society was always on the periphery, always close to home.

I was just nineteen or twenty years old when all of my friends started going into the service to fight in Vietnam. I decided I didn't want to wait to be drafted. I was just going to go ahead and sign up for military service.

I made my way to downtown Manhattan to the Army induction center at 39 Whitehall Street. This was a time when the military was so eager for soldiers they were taking people who had one leg shorter than the other.

I was determined to enlist. But when I went to take the physical, I got quite a shock. I had always been a healthy kid and never really encountered medical problems, but the doctors conducting the physical rejected me because they said I had a heart murmur.

I never knew I had any problem at all. I wasn't going to let it stop me, either. I decided to go to a different doctor out in Westbury, Long Island, a town not too far from where I lived. I remember telling the doctor that I wanted to get a physical and instructing him that I wanted him to make it clear that my heart was okay because of the military service.

I think he caught kind of an attitude at first. He seemed to think I was trying to get out of military service. But it wasn't too long before he realized what I meant. When I gave him the papers from the doctors at Whitehall Street, he understood. But he ended up confirming that I did indeed have a heart murmur. It's something that never bothered me or gave me any symptoms or problems.

There were riots. My friends were going to war and returning home to be buried. I was bitter. What was I here for? I tried so hard, twice, to go into the service. Being unable to join up left me drifting a bit. I was in a store once, and looked outside and suddenly saw all of these people who had come into the neighborhood—they had helmets, shields and billy clubs. It felt like they were invading our community and they left a feeling of intimidation in their wake. It was a time of fear and hostility.

I was no different than any other nineteen-year-old kid who had little direction. There were times when I would come to a situation and found myself at a veritable fork in the road. Whether it was a situation that came up while I was out on the street or hanging out with my friends, it was so tempting to do the wrong thing, the violent thing and the passionate thing. There were times when I would approach those moments, be in them, and was so close to making that turn into doing something awful. I can see how people with little direction, no family, no grounding or foundation in their lives could be lost to the streets.

How many people were lost because of stupid violence and a lack of guidance? Another generation just gone, an unnecessary loss of potential.

In hindsight, not being able to enlist was a blessing in many ways. I still have my draft card, and I look at that and remember how much I really wanted to go. But I believe there was a greater reason things turned out the way they did. I was so hostile and volatile at that time; I probably would have gone overseas and done something stupid. When you are nineteen, you do silly things.

But military service was not in my future, and after trying to sign up twice, I gave up on the idea and decided to just focus on work.

My family was forced to relocate again, this time to a place in Westbury, and I moved back in for a short time. I was still working at Nobman's, but I had finally bought a car—my first car. A red 1954 Mercury, a beautiful car with big chrome bumpers. The only catch was: It was an automatic.

In those days, an automatic just wasn't cool. A stick shift, that was cool. I always regretted that I didn't have a stick shift. So being mechanically inclined, the idea occurred to me: Why not make my car into a stick shift? Why not—it should be doable, shouldn't it?

So I did my research and was feeling pretty confident in what I could do. We didn't have the computer then, so research consisted of a trip to the junkyard, where I found a 1954 Ford standard shift transmission, along with all the parts and materials I needed, including a clutch pedal.

I brought everything I needed over to a friend's place, and he helped me get the work done. As we went along, I looked at the heavy chrome bumpers that were all the rage at the time and decided, *Who needs them?* I took them off.

I drove it in an automatic and, a mere five hours later, drove it out a stick shift.

Now I had some mechanical ability and a can-do attitude, but I was no expert. I remember driving down the parkway, listening very closely for things that might be going wrong.

At that time, Long Island was less developed, and a drive on the parkways was a little more desolate than the trip through crowded suburbs that it is today.

I had the radio on but all of a sudden there was a thump. I turned the radio off and heard "thump, thump." I didn't know what that was, so I turned the radio up and made it louder to drown out the noise.

Well, that didn't solve the problem. Finally, I pulled over to the side of the road, my heart beating while I prayed that nothing was seriously wrong. I was so relieved to see it was only a flat! I confidently went to get the jack so I could start changing it when I stopped short. That's when I realized that, back then, there was a reason for those big heavy bumpers, that I had removed from my car: the commonly used jack at the time used the bumper to help lift the car off the ground. So now I was stuck until, finally, someone else came along who had a jack that fit underneath the car, and I was able to use it to fix the tire and get on my way.

Despite adventures like these, it wasn't long before I realized it was time for me to get my own place. So I went out and found an apartment.

It was an efficiency apartment in Hempstead, and when I say "efficiency," boy, do I mean it! This second-floor place had a tiny bathroom, a little kitchenette and one main room. That's it. I had already signed the lease and paid the landlord so I could move in when it occurred to me that I should probably say something to my parents.

It wasn't so much a question of getting permission but one of respect. So I took my father aside one day and told him that I thought it was time for me to move out on my own. I asked him to come look at a place I was thinking about renting. I drove him over to the place that I had actually already rented and asked his opinion. He looked around and seemed to approve. "That's nice.

That's good," he said. It may have been unnecessary, but it made me feel better and I'm sure that, deep down, my father appreciated it as well.

It was around 1965 that I went to work for Grumman, which was based in Bethpage, another Long Island town. For decades Grumman was one of the largest employers on Long Island and played a significant role in the aviation and space industry.

Grumman was one of the contractors whose employees became a critical part of the Apollo program, and while I still carried that same confidence as I did all those years before when I went to ask Mr. Nobman for a job, I was lacking in some of the technical skills I needed. I really think it was partly luck and the company's need for a warm body that got me my entry-level job as a wireman technician.

But Grumman was building the vehicle that was going to send a man to the moon and it was work that drew in the whole company. I had learned some very basic mechanical and electrical work at Nobman's, which had a lot of different departments, but learning how to repair lamps and lawnmowers was not proper preparation for the work Grumman needed done. I took a lot of technical courses to round out my skills and help me progress.

I found at Grumman the guidance and chances to succeed that I wish now I had been able to find at school. It was a place that if you applied yourself, the work and effort would be noticed. I was given an opportunity to grow, to progress and to have someone take notice of me. My potential was recognized and I was pointed in the right direction—and that in and of itself was a blessing.

It was a different time in the life of corporate America as well, and Grumman was a great place to work. The thing I always remember was that, each year at Christmas, it was almost always cold and snowy, but quite festive. Every year the workers would get a bonus and a frozen turkey.

I'm sure in later years that policy changed and people simply got coupons for the turkey. Those bonuses were probably ended after a time, too. But when I first started there, there was an effort to care for the workers that is missing from many companies today.

The work and the opportunity I found at Grumman were like giving sunlight and water to a starving plant. I was very enthusiastic during this time and was always looking for a chance to further my station in life. I applied to and was accepted in every technical training school they offered. I was always looking for work, even though Grumman wasn't my only job.

After working a full shift at Grumman, I would head over to the Sinclair Gas Station in Westbury, where the keys would be handed over to me for the 6 p.m. to 10 p.m. shift. It was hard to work two jobs but it wasn't like I had anything else to really do, as my social life wasn't exactly thriving.

But I was young and had a lot of energy, and it was a busy time at Grumman, where my real future was. One of my co-workers, a man named Tom Tooley, began encouraging me to take advantage of something called temporary duty.

Temporary duty was work that would take me away from Grumman's headquarters in Bethpage and place me on an assignment for six to eight weeks at a time. It must have been tough for some of the older workers with spouses and young children to support.

But I had no such responsibilities, and even though Long Island was my home I wasn't tied to any particular location. I was younger and could explore opportunities to travel, and as I continued to work at Grumman and expand my skills, I started to feel that I had more talent. I felt I was being underutilized and that I could do more.

I was once told that in order to stand out in a crowd, you have to stick your neck out. So that's what I started to do. I went to several company schools to train on the equipment I would be required to maintain on my new assignment.

I was selected and got my first temporary duty assignment at the Naval Air Station North Island, San Diego, California, working on the *USS Kitty Hawk*, which was an aircraft carrier. I'm sure I mentioned it to my family, but certainly not in any way that involved seeking their permission. There weren't any long discussions about the pros and cons. I had left the nest already, so to speak, and, in those days, once you spread your wings, you were out on your own.

So I was off, heading across the country for this new assignment. It was a job that required me to get security clearances and to handle a lot of information all at once. When I arrived, I hooked up with a few more technicians at that location to get my bearings, but the first time I saw that aircraft carrier, I was overwhelmed.

It was a huge ship. My job involved making sure I was to work on and provide support for all the ground equipment the Navy would use to support its planes—specifically, the A-6 Intruder, an attack aircraft, and the E-1 Tracer, used as a kind of scouting aircraft. My security clearance for that assignment allowed me access to all of the fleet's ground activity.

But it wasn't the work that was so intimidating. Just getting onto the ship was a new experience.

For starters, I had to learn the U.S. Navy protocol for proper ingress and egress of equipment onto a ship.

That wasn't all. My security clearance, I was told, gave me the privilege of using the ship's forward entry ramp to go aboard. This ramp was flanked by a U.S. Marine in uniform, holding a huge rifle.

There were 5,500 people aboard the USS *Kitty Hawk*, making it somewhat like a small city. Most of the sailors, who always had to turn and salute the flag before getting onto the ship, used the aft, or rear, ramp.

I usually used the aft ramp. I felt more comfortable and less self-conscious, more like one of the regular folks, entering the ship that way.

But I wasn't there for fun. It was crucial that I completed my assignment before the ship left port for its mission off the coast of Vietnam. I didn't relish the thought of being a part of such a trip, so I worked diligently.

Still, everybody needs to eat and I usually ate in the ship's cafeteria. That's where most of the ship's inhabitants had their meals. One couldn't imagine a noisier scene than that mess hall, with people loudly chattering together as they sat and ate at long tables, with others shouting orders or clanging lunch trays.

One day I decided to get my courage up and do something a little different. I told myself it would be nice to use the forward ramp and eat just once not in the mess hall, but over where the officers dined.

Slowly, I approached that forward ramp, and in my heart I was sort of hoping that the sailors behind me wouldn't see. When I finally got to the top, the marine standing guard snapped to attention before the officer of the deck checked my credential and let me continue on.

Then I was in, and made my way over to the officers' meal area.

This was no mess hall. What a drastic difference!

This was a real restaurant: white linen tablecloths, carpet on the floor, servers with cloths on the arms of their jackets—not to mention the baby grand piano.

As I took my seat and tried to decide what to eat, I marveled at the luxury and formality of this new environment. I could only wonder if the sailors knew whether this place even existed and I can remember thinking that, surely, there would be mutiny if they found out.

It wasn't long before my work was finished and the assignment came to an end. With that, I made my return to Grumman's headquarters in Bethpage, and to home. But a single assignment like that wasn't enough to satisfy my restlessness. I found myself applying again for a temporary duty assignment, and got a few more. I would spend several weeks at a time working on aircraft carriers like the *Kitty Hawk*, including the *USS Constellation* and the *USS Forrestal*. The *USS Kitty Hawk* was one of the greatest aircraft carriers that has ever sailed, and has served her country for almost fifty years before being decommissioned on September 2, 2008. My first experience was on the *Kitty Hawk*, and it will always be my fondest.

As the months went by and I was successful on my temporary assignments, people started to persuade me toward a higher goal, bigger than your typical temporary duty assignment. I was encouraged to go for a one-year assignment and my confidence was high—or maybe I was just a little cocky—so I went ahead and put in my application.

This move put me squarely in the middle of one of the company's most significant projects. Grumman had been chosen as one of several subcontractors to do critical work for the United States space program, including designing, building and testing the lunar module, the vehicle astronauts used to go into space and land on the moon.

To train for a potential one-year assignment, I began

working on the real McCoy—the vehicles that were destined for outer space.

It was like working inside a real-life science fiction movie. We worked in a "clean room" and were dressed in white from heat to toe, with white jumpsuits, caps and gloves. I was up close and personal with several vehicles, including LM-5—the lunar module that would come to affectionately be known as "The Eagle," which took our first explorers to the moon.

My training was to prepare me to replace one of the technicians at the Kennedy Space Center in Cape Canaveral. I spent months getting to know the equipment, learning the ropes, getting myself ready. I remember being about to leave for the day, looking back over my shoulder at this incredible vehicle, and feeling a chill go up and down my spine as I marveled at how this machine that I had been working on would be heading to the moon.

It was around this time that my usually dormant social life got a bit of a spark.

I wasn't much of a dancer, but I could fake it pretty well. I had a strategy: I'd wait until a song was half over, and then make my way to the middle of the crowded dance floor and do my thing. It always seemed to work, and I'd get by just fine.

One night my friend Carlos James, from Oyster Bay, and I decided to go out for a night on the town in the big city. We left Long Island and drove into Manhattan, managing to find a parking spot before going into this club called Ché José.

I don't know what I was thinking. I couldn't salsa dance. I couldn't speak Spanish. I guess it just sounded like a good idea to go to this club. Little did I know it would turn out to be one of the best ideas of my life.

Carlos and I went up to the door of Ché José brimming with confidence, and were immediately turned away—just because we weren't wearing sport jackets.

Well, that didn't sit well with me. It bugged me to be shut out for something as silly as not meeting the proper dress code. I couldn't help but think, *No sports coat? Who do they think they are?* But, after all, this was the 1960s. That's the way some things were.

For whatever reason, I just couldn't let it go. Several weeks later I decided to give it another go. I dug out my only jacket and made my way back to the club on a Friday night. This time I was granted entry.

Right away I was in a bit over my head. I liked the music, even though I couldn't understand the lyrics. It was a typical club, dimly lit and a bit smoky. Now, salsa songs can go on forever, and it was pretty obvious that I didn't have a clue as to what I was doing, but I thought I was getting along all right. There I was, moving around and doing my best salsa dancing with a girl at the club. That is, until the song ended and the girl I was dancing with realized that I had no idea how to salsa dance. As soon as I turned my head, she was gone.

I didn't even know who that girl was; she was just someone with whom I had started dancing. But the fact that she left me alone on the dance floor didn't mean a thing because standing in front of me was another girl—the most beautiful girl I had ever seen.

I was stricken. I couldn't help myself. Immediately I swallowed my nerves, turned on my sophisticated personality and approached this gorgeous girl.

I think she felt bad for me because of my lousy salsa dancing. She was quite nonchalant as I asked if I could sit near her. "Well, suit yourself," she said. I remember she had a cigarette in her hand and it looked eight inches long. As I watched, I could tell she wasn't actually smoking it, but it was the cool thing to do in those days, to have a rum and Coke and a cigarette. So she was just playing along.

She started talking to me and I think her pity for me kept her guard down. Her name was Amina Suddin, and we talked for a little while as best we could through the loud music before I got my courage up and asked for her phone number.

Amina appeared to hesitate a bit, but then she gave me her number. I started to write it down on the only thing I had available, a wet napkin, hoping I hadn't swapped any of the numbers or misheard her through the blaring salsa music.

We said our goodbyes that night and I promised to call. I wondered what kind of chance Amina would give me, but held onto my hopes and the feeling she'd left in my heart.

Of course, I couldn't call right away. Since that was a Friday I decided I had to wait a couple of days. I called at noon that Sunday. Her sister Lillian, who was sometimes called Ronnie and who was to become one of my favorite of Amina's sisters, answered, and she told me: "Hold on, I think she's sleeping."

Now, for only a second, I thought, *Sleeping? What's up with that?* But I quickly came to my senses as I waited for Amina to get to the phone.

After a few moments, however, Amina did pick up the line.

"Amina, this is David," I said.

An instant of silence followed my greeting. There it was again, that hesitation.

"David?"

Amina repeated my name as if it was the first time she had heard it. I gulped and prepared myself for rejection. It's not like I hadn't expected it. But then, like a ray of sunshine, she said, "How are you?"

I knew then that I had a chance. Things were looking up for my once-nonexistent social life, and Amina and I started seeing each other.

We dated for two wonderful months before I got the news that would help change the course of my life.

Grumman was changing the assignment I had planned for. Now I was being sent to Houston, Texas, and replacing a technician working on the lunar test module. The assignment included a one-year contract, and the company would pay for my move and a one-bedroom apartment, including utilities. They would also fly me home every three months.

This should have been good news. This is what I had trained for, and this was one of the company's most important projects. I had pursued this opportunity by my own choice, and worked hard for it.

I didn't want to show it, but I was actually more disappointed than anything else. After all, I had just met the girl of my dreams.

And Houston! I was from New York. I grew up in the backyard of one of the world's biggest, most sophisticated cities. When I thought of Texas, I had visions of dirt roads and cowboys, of a backwards kind of living. How would I survive in a place like that?

There were other concerns as well. It was the late 1960s, and I was a young man who was half-black, being sent by myself, thousands of miles from family and friends and everything familiar to me, into the deep South.

While I was not immune to racism in New York, it was impossible to ignore the turmoil that had engulfed the nation, especially in the southern United States. There were race riots in various cities and heated battles for civil rights that led to ugly scenes that still resonate in our history.

It was about this time that Alabama's notorious governor, George Wallace, who had gained attention by standing in the doorways of public school buildings to keep black children from attending and was known for the phrase "segregation now, segregation tomorrow, segregation forever," ran for president.

It wasn't until 1967 that the U.S. Supreme Court struck down a Virginia law that banned interracial marriage. That year, Thurgood Marshall was the court's first black justice. In 1968, Martin Luther King, Jr. was assassinated, and more riots followed that horrible tragedy.

The turbulence was far-reaching. Robert Kennedy had been killed. The North Koreans were creating international unrest while Vietnam was still consuming the lives of American soldiers.

And there I was, just a kid in my twenties, going off into the great unknown.

Despite these concerns, I knew there was no turning back. I had made a commitment I couldn't break, and didn't really want to. I still had my ambitions for being a part of something important at Grumman. My only real regret was the bad timing of this move in terms of my relationship with Amina.

I put every positive twist I could think of into explaining the situation to Amina, to tell her in a reassuring way why I had to take this assignment. To my surprise, she was quite encouraging. She knew I needed to stay focused and needed as much support as I could get.

And so with her blessing, I signed the contract for my one-year assignment and got ready to move my belongings across the country. It was then that I realized I had virtually nothing to move, since you can't really pack all that much into a small efficiency apartment in Hempstead, a Long Island community not far from Grumman's headquarters.

I promised Amina we would be together. I packed up my 1968 Cutlass Oldsmobile and went off toward what was then known as the Manned Spacecraft Center in Houston.

PART II

I had never before driven more than fifty miles at one time; as a matter of fact, just getting out of New York City was an interesting adventure. But I tend to thrive under pressure, so this was a good challenge for me with which to start the trip.

I have never really paid too much attention to discrimination before. I wasn't ignorant of it, but maybe because in my experience the signs had been subtle, or maybe it was being biracial—I just didn't get it. But driving through the southeastern half of the U.S. made me realize the cold rejection one can receive based solely on their appearance.

The trip by itself was enough of a challenge, with my New York license plates attracting plenty of attention, and long, lonely hours of driving in unfamiliar territory. It was jarring to see George Wallace's campaign posters as frequently as I did. After ten hours of driving, I thought one could just pull into a hotel, register and then get up the next day and keep going. I did have a deadline for getting to Houston, so it was important to make good time on the drive.

I approached one hotel; I remember there were only seven or eight cars in the parking lot. I had been driving all day and I was just exhausted. All I wanted was a place to rest.

Going up to the counter, I asked for a room for the night.

The person behind the desk took one look at my license plate, and one look at me, and declared, "We are full for the night."

I couldn't believe it. I stood there for a few minutes, not knowing what to do. It was obvious the hotel wasn't full, yet I was denied access. I couldn't find a word to express how I felt through my fatigue, as it dawned on me that I was being turned away because of the color of my skin and where I was coming from. That's when I realized this was going to be a very long trip.

It was that kind of episode that gave me renewed sympathy for my father, as it gave me a small, subtle taste of what he must have gone through his entire life.

Later, I was able to get a room in a different hotel, where I showered and rested, thinking of the distance I had already traveled and what was still ahead of me before drifting off into a deep sleep.

The next morning I grabbed a cup of coffee and as I got in the car, I had managed to slightly bump my rearview mirror. I looked up, stopped and stared into that mirror. What greeted me was the sight of everything I owned in the back seat and trunk of my car.

That's when it hit me. That's when I knew that, no matter what the contract I had signed said, this was a one-way trip.

I had left my life behind in New York: my family, my girlfriend, everything I had ever known. I felt so alone. All I could do was sit in the driver's seat as thought after thought popped into my head. What if something happened? What if this was the wrong choice for me? What if it didn't work out? What if…?

There was no room for that. If I were going to let questions like that stop me, I never would have left New York. I decided to push those thoughts out of my mind and keep going. I had to meet my report date in Houston the next day.

As I drove into Houston, I got my first pleasant surprise: no dirt roads. There were no cowboys that I could see. The roads were paved, the city looked quite modern. I shook my head and laughed at myself. What was I thinking? Well, my thoughts were probably no different than that of a person who would be going from Houston to New York, and probably anticipated getting mugged as soon as they got out of the airport.

That introduction to a new city, that shattering of my assumptions, taught me to take the blinders off and to never be judgmental.

My arrival in Houston, at the Manned Spacecraft Center, was like being ushered into a new world. To be sure, this was a critical time for the space program, with the country racing to beat the Russians and the specter of what came to be known as the Apollo 1 disaster in everyone's minds.

The tragedy had claimed the lives of three Apollo astronauts on January 27, 1967, at Cape Canaveral, where I was originally supposed to be stationed on my yearlong assignment. The Apollo/204 mission, which was the original name for what became known as Apollo 1, was supposed to send a manned command service module with three astronauts into the earth's orbit. The command service module, or CSM, was a companion vehicle to the lunar module I had spent so much time working on.

Gus Grissom, Ed White and Roger B. Chaffee were aboard the command service module for a training exercise that day when the module, which sat on a Saturn 1B rocket, burst into flames, killing the men. It happened so fast—the crew was breathing 100 percent pure oxygen, and not everything in the cockpit was fire retardant, not to mention the hatches that opened inward and made it tougher for them to get out on their own. Temperatures inside reached approximately 1,400 degrees, and the crew died within seconds.

With the political implications of NASA's missions hanging over everyone's heads, and the Apollo 1 catastrophe heavy on everyone's hearts, security at the Houston facility was tight, and testing of the equipment was treated with the utmost seriousness.

I reported to a man named George McNally, who in turn reported to a man named Howard Hollin. They helped finish my paperwork and fill out the required information so I could get a security clearance to enter the NASA facility.

It may have been overkill at the time, but everything was treated with the utmost security and a U.S. government secret clearance was required. I remember driving through the gates of the Space Center for the first time and reporting to work. It was intimidating, exhilarating and nerve-wracking all at once.

Meeting my peers, all highly trained spacecraft technicians, made me catch my breath. That's when you suddenly realize you're the new kid on the block and you have to prove your capabilities and earn your stripes.

It took time, but slowly I got to meet everyone, and they welcomed me with open arms. My presence was a much-needed personnel boost since the technicians were working six to seven days a week, days that were ten to twelve hours long. Any kind of assistance was welcome and it didn't take me long to become a part of the community.

I was assigned to the avionics division.

At the time there were several lunar test articles, or LTAs, being tested by Grumman at various sites around the country. These LTAs were critical for helping to work out problems with the equipment that would eventually send men to the moon. In Houston, LTA-8 was being put through the ringer for thermal-vacuum testing, which was crucial for the safety of the astronauts, while other LTAs were being examined for the effects of structural shakings, vibrations and various other systems evaluations.

I was stationed with a headset, in front of this ground control power equipment that was so lit up it looked like a Christmas tree. This equipment provided the power for the actual lunar module testing. The chatter coming through the headset was unbelievable; it seemed there were always fifteen-conversations going on at once. But this is what my training was for, so it was no big deal.

The trick was to only respond to the test conductor when you heard your station's call sign in your headset, and then perform your sequence. It meant there was no room for error. It was not unlike learning to play the drums in high school, when I had to concentrate and know exactly when to join the music. At the Space Center, one had to stay focused, listen carefully and, most importantly, stay awake and be ready to do your part.

When I arrived in Houston, the technicians had been working long hours with plenty of overtime. I wanted to work as much overtime as possible, even on weekends. I mean, what did it matter? With Amina back in New York, what else did I have to do?

My social life was nonexistent and everyone I knew was back home, so I had no choice really but to focus on my work. And since I was proficient enough, I was able to pitch in and get the hours I needed while giving some time off to several of the guys, who maybe had families at home they were eager to see.

I worked every shift imaginable, which took some getting used to but after a few months of settling in, I got used to it all: the area and the people. I never caused any problems with people and worked hard, but the shift work was starting to take its toll.

I worked so much and racked up so much overtime that eventually I was considered when an opportunity opened up for a higher post. At this point I had worked my way into a supervisory position, and even though there were guys who had seniority over me and I figured I didn't have much of a shot, I was given a chance.

No matter what anyone says, if you're in a group and you're suddenly viewed as the boss, you're looked at differently—and I was. Even the people to whom I was closest began to pull away. That was tough to go through, but I did okay and could hold my own among the technicians.

The work usually wasn't glamorous. However, we were all keenly aware of the importance of what we were doing—after all, many of these tests were pre-Apollo 11, so they provided information that was critical to the overall success of the space program.

But to be honest, it was sort of boring. For ten to twelve hours at a time, you might be standing inside the LM's cabin (since there were no seats), with your only connection to the outside world coming through the headset. I and the other technicians doing the testing would stand, activating various switches and dials and giving readouts. It was tedious. But you always felt good in the cabin of the LM—and much better when your shift was over!

There were other technicians more experienced than me. Jim Travlos and Al Fischer were the best. Whenever I was in the cabin and was asked to perform a task, I would always hesitate a second just to be sure I was performing the proper sequence of events. Those guys never hesitated, and Al Fischer was the best of the best.

It was critical for everyone to pay attention, not just

because we were dedicated to making sure we got it right. Once something went wrong during the testing, it would set off an alarm. And after the first time you had a master alarm go off and the lights started to flash, it got your attention. After a while you got used to it, given the number of tests being performed.

It was indicative of the times, and the sense of urgency and importance that surrounded the space program, that whenever you needed additional equipment or resources, you got it. There was no constraint. Whatever you needed to complete the test was sacred.

I spent so many hours inside the lunar module, and I think I activated just about every button and switch. That moment of wonder I had at the end of one of my shifts back in Bethpage, that slight questioning of whether this vehicle, which had walls so thin you could actually take a pencil and poke a hole through them, could really bring men to the moon, was eventually erased.

That small moment of doubt was singular. I never had a moment like that again. My confidence grew each day and every day that I worked on the lunar module.

Much of the lunar module testing was done in the space environment simulation laboratory (SESL), which was in building 32 and building 49 vibration and acoustic lab of the Manned Spacecraft Center. The SESL had two separate sections, known as chamber A and chamber B, and was one of the world's largest thermal vacuum chambers.

This laboratory was critical to the success of the space program because it allowed NASA to conduct a wide variety of critical testing in a space-like environment and to check life-sized equipment such as the actual lunar module for design and development issues.

Chamber A was where they put the command service

module, while the lunar module would be studied in chamber B, which was a bit smaller. The command service module and the LM were companion vehicles, both intended to bring humans to outer space.

Chamber B was used to conduct vacuum tests on the lunar module, and it had a removable ceiling so that the LM could be dropped inside. NASA had different testing facilities to examine the effects of extreme cold and the effects of extreme heat as well. The vacuum tests were supposed to simulate the harsh, airless environment of space, and once the environment was created, it carried with it all the risks and dangers of being in space. They would use a fifty-ton crane to carefully remove the top portion of chamber B, and then carefully place the LM inside. A vacuum was then used to suck all the air out of the chamber—a procedure that by itself took eight to ten hours.

By the time they were done, there would only be enough air left in the chamber to fill a ping-pong ball. As one of the technicians working on the LM, I would do whatever it was that needed to be done, and then leave, turning things over to astronaut Jim Irwin, who would be in the vehicle during the vacuum test as the commander.

Later, Irwin would become the eighth person to walk on the moon and piloted the lunar module during Apollo 15, which launched July 26, 1971. It was America's fourth lunar landing. Other crewmembers included David Scott, the commander, and Alfred Worden, the command module pilot.

Jim Irwin was also one of the first people to drive a car on the moon, called the lunar roving vehicle or "moon buggy." On earth, the LRV moon buggy weighed an impressive 480 pounds, but on the surface of the moon, it was only around eighty pounds. It was a strong vehicle that could carry twice its weight—sturdy, but not very fast. Eight miles per hour was the top speed the moon buggy could hit. But it wasn't built for speed, and would set you back around $9,500,000.

Think about that! You're going on a long trip and you take your own ride along as part of your luggage. Once you arrive, you unpack, get in and drive away. How cool is that? Only two other Apollo flights, Apollo 16 and 17, would have that same luxury.

It was typical for the technicians to work closely with the astronauts, and an exciting experience. It was a stark reminder of how important it was for us to give our best efforts and to stay 100 percent focused on our work—because they would be counting on our commitment and desire to get it right.

And the job came with its own perks, at least for me. One day I was going along, minding my business, when I got on an elevator with Alan Shepard, the first American in space and the fifth man to walk on the moon.

I was overwhelmed. This was an Apollo celebrity if there ever was one, standing next to me! I kept my cool. Alan had an engaging personality and we made small talk for our few minutes in the elevator before we each went our separate ways. The whole memory is kind of hazy—I guess I was starstruck.

In April 1968, it was decided that testing of the lunar module would begin and it would be ongoing for several months. As I described it to my friends, the lunar module in chamber B looked almost as if it were a person in the intensive care unit. The LM was hooked up to so many instrumentation devices and electronics, with wires and hoses all over it, that it wasn't very pretty. But that's exactly how it looked in the chamber, and all those sensors and wires were needed in order to evaluate the results of the tests being performed.

Varying data points had to be monitored, that way the engineering staff would have the proper information for all systems testing so they could evaluate the results. Most of my time was spent in support of the testing going on in chamber B and it was mostly normal testing, with the usual spells of boredom.

It was necessary to have more than one lunar module vehicle for testing and planning purposes. The original plan was to have LM 1 and LM 2 be flight-worthy and flight-tested, but after the first successful, unmanned LM 1 flight during Apollo 5 on January 22, 1968, it was decided a second flight test was not necessary.

There was additional testing that gave greater confidence in the lunar module's ability to withstand landing on the moon. After all, LM 2 was a flight-worthy vehicle and all of the sub-systems were working exactly like the vehicle that would take us to the moon.

Later that year, the Apollo program had a great success. It was December 1968 when astronauts Frank Borman, James Lovell and William Anders became the first humans to see the far side of the moon for themselves. It was also the first manned launch of a Saturn V. Apollo 8 also became the first human spaceflight mission to reach speeds that allowed it to leave the gravitational influence of the earth and then orbit another celestial body.

Before they launched to go to the moon, I had a chance to meet Frank Borman. It was thrilling. I got his autograph. I remember talking with him about the upcoming mission and how special it was and all the important firsts it would achieve. I believe Apollo 8 was a huge step toward America closing the gap with Russia, and it was an incredibly significant flight. And I got to shake hands with the man who led the mission and made history.

I worked in building 36, where I would spend countless

hours in support of the LM 2 drop test, which included dropping the lunar module to test how the vehicle would stand the impact of a landing.

LM 2 was chosen for the drop tests because it was a truly flight-worthy vehicle, and the alternative, LM 3, didn't contain all of the operational subsystems such as plumbing. There was a concern that doing such tests with the LM 3 would be pointless because it would not truly represent a flight-worthy vehicle.

But using the LM 2 meant that some questions could be almost definitively answered, including how an actual flight-worthy vehicle would land, how it would stage and how well the pyrotechnic devices would operate.

In May 1968, several drop tests would be done to make sure that all the systems on the lunar module would perform as expected. The LM that would go to the moon had lunar sensing probes, which consisted of an aluminum pole approximately sixty inches long and covered in a gold film called kapton, which was useful because it could tolerate a wide variety of temperatures.

Each of the lunar sensing probes had a switch-activated mechanism that would lower the pole to the moon's surface. Once the pole touched, the lunar surface sensors would let the crew know via a large light that they had achieved physical contact with the moon.

Usually there would be three lunar sensor probes attached to the descent landing legs. They were mushroom-looking pads on the legs; however, there was no sensing pole attached to the front landing gear. This was so that the crew would not be impeded while descending the front ladder, or face the risk of tripping while on the surface of the moon.

During the drop test, the LM 2 would be placed at approximately the same distance as if it were making its final approach for a descent to the moon, but mathematical calculations were made to duplicate the actual descent.

The other drop test was to make sure that the pyrotechnics were working as designed, as well as the physical separation of the two stages: the ascent stage from the descent stage. On first thought, perhaps the idea of miniature fireworks being exploded during the delicate lift off and landing of the lunar module sounds strange. But pyrotechnic devices played a major role in the Apollo program.

Approximately 200 of these devices were used in every Apollo flight, with no failures. They were actually explosive nuts and bolts used to deploy a mechanism, or separate two sections of the module. They had a very reliable outcome. The failure rate was once estimated at approximately one in 50,000 and they were used in various areas, landing gear operation, ascent/descent, pressuring of the valves, etc. We provided technical support for the drop tests, which were being performed in building 49.

Certainly this testing wasn't glamorous. But it wasn't without its dangers, either. It was mandatory around this time that every technician who worked in or had access to these chambers had to take particular safety training and go through a certification process.

Part of that training included information about hypergolic fuels, their toxicity and the hazards of being exposed to such substances. Hypergolic fuels usually consist of two different substances that ignite instantly when they come into contact with each other. And while these fuels were handled with great care and usually kept under strict control, they were highly toxic and were not anything around which anyone would choose to be.

I had no direct contact with the hypergolic fuel used in the lunar module, but I was close enough. I couldn't help but be concerned with how poisonous these fuels could be, and just the thought that the danger could be so close was dreadful.

The drop testing went on for quite a few months, and for the most part things were uneventful. But as the weeks and days flew by and the space program geared up for its biggest show—trying to send a man to the moon—I was treated to the scare of my life.

The final drop testing was scheduled for May of 1969, and everything was proceeding as planned. During these tests, the lunar module was to undergo the same processes it would go through as if it were landing on the moon, with its various segments separating during the takeoff and landing routines.

By now, I, along with many others, was working long days and many hours of overtime to support this effort. Things were going along as usual on this particular day, and there was just one other technician, a man named Gerry Grevsted, with me in the LM.

During our assignment, I was to act as the "commander" and Gerry was playing the role of the LM pilot. We were standing in the cockpit, and it was a typical eight- to ten-hour shift. Our job was to go through every switch and dial and make sure that things were configured as if we were the astronauts going through the actual trip to outer space.

Since LM 2 was a flight-worthy vehicle, the tests were performed in real time. I have done many system tests like this, and probably could have put the switches and dials in their proper positions even in my sleep.

We were suited up, wearing our white jumpsuits, with our headsets linking us to the outside world. The blinking lights and complex instrument panels dominated the cramped quarters in which we were working.

About ten or fifteen other people were sharing the same line that we were plugged into via our headsets, though the majority of technicians were in other locations, performing tests outside of building 49, where we were. But the chatter coming from the headset wasn't enough to offset the loneliness that sometimes made the testing dreary.

It was just Gerry and I, taking direction from the test conductor, who had us activating different circuit breakers and switches, and then giving readouts of the various results.

The cockpit was dimly lit to make it easier for the astronauts to read the gauges and equipment. Everything seemed to be quieting down as other tests were done and other subsystems came online.

Things were peaceful, routine, as we chatted with the test conductor. Then we were rocked by a sudden, shocking explosion that shattered the cocoon-like calm of the lunar module.

It sounded like a bomb had gone off, and the paper-thin walls of the LM transmitted the vibration easily.

The master alarm began to blare as the instrument panels on the lunar module appeared to go haywire, instantaneously flashing and giving random, anonymous readings.

Terror and uncertainty shot through me. I was standing in as the commander of the "mission," and it was up to me whether or not we would bail out. Yes, in reality, this was simply a test—all we had to do was open the entry hatch and evacuate.

We weren't in space, but as technicians we were trained to handle the tests with the utmost seriousness, and while something had obviously gone wrong and we didn't know what it was, it was crucial to get as much information from this situation as we could. Better to have a mistake of any kind occur while the LM was still on earth, and not in the deadly, alien environment of outer space, such as what took place on Apollo 13.

My pulse was doing double time; I had to make the call to the test conductor and seconds were slipping by. Just as I was about to open my mouth, the test conductor realized we were having a serious problem but no one was sure what had happened or what we should do next.

They kept asking us for readouts and pressure readings on some of the gauges and I was only a millisecond away from leaving the cabin of the vehicle when I paused and remembered to stay calm and not to overreact.

The only thing in my mind was my hypergolic fuel training and the deadly consequences of making a mistake around those volatile substances. We were doing tests in real time, going through each step as if it were really happening—so did that mean that the LM 2 that was flight-worthy, the one we were in, had ever been loaded with this fuel?

My imagination was running wild. No one had any answers and no one was talking to me about what had happened. The seconds and minutes ticked by but time inside the LM felt endless.

Eventually, things started to settle down as we came to understand that, whatever it was that had happened, there was not any apparent, immediate danger.

Later, it was determined that the problem was in the oxidizer tank on the ascent stage of the LM 2, as a vacuum was being pulled on the tank to eliminate any possibility of vapors prior to the drop test. Apparently excess vacuum pressure caused the tank to implode. It was this implosion that created the bomb-like noise.

I have never been more frightened in my life, but I'm also glad that I never made the call for Gerry and me to leave the vehicle. In the end, the drop test was delayed for a short time, and then finally it was successfully completed.

The experience made me think of our brave Apollo astronauts, who risked their lives in space. If something happened to them in their space vehicle, the options were very limited—they were truly brave and courageous heroes. Everyone in America owes them a great deal of respect. We would never have accomplished the magnitude of what we did without them.

Houston was one of the places where we handled the testing and configuration of the LM and the CSM. But nothing launched out of Houston. All of the missions launched out of Cape Canaveral. It had to do with taking advantage of the earth's rotation; sending the missions out of Cape Canaveral on the East Coast meant the vehicle could add 1,000 miles per hour of velocity to the spacecraft.

So in Cape Canaveral, it was their job to assemble the vehicles, do some testing and then handle the launch. Houston would usually take over after the Saturn V cleared the tower.

After the drop tests were finished, LM 2 had to be reconfigured. It was getting ready to ship. The vehicle was going to go on display at the Smithsonian Institute in Washington, D.C.

There were a lot of things that needed to be done so the vehicle would be in show-worthy condition, including pieces that needed to be reworked and propped up so it could be displayed properly.

Reconfiguring the LM was a joint effort among several technicians, including ones like me downstairs in avionics and others like my co-worker Richard Specion, who worked upstairs in the instrumentation division.

Ritchie and I worked closely together. He was a young guy and one of the sharpest knives in the drawer. We were shocked when, sometime later, he died suddenly of a heart attack. I know his family and friends are proud of the contributions he made to our space program.

I'm not sure what my motivations were, with my emotions running high and the scare I had inside the LM, but I felt a strong attachment to this lunar module. With that on my mind, I took a magic marker and, when no one was looking, reached up into the descent stage engine plume. I wrote my name there and his was placed alongside mine. Now, I have confessed another one of my sins, and only hope that after forty-one years, the statute of limitations for defacing government property has expired.

But no one found out about my crime, and life went on.

Even though there were thousands of other workers on the space program, I was lonely and tired of living through nothing but my job.

Thankfully for me, Amina decided without any hesitation to join me in Houston. She had been working in an administrative position for a New York law firm at the time, but chose to leave her job to join me in Texas. It was the start of our new life together and the end of my lonely days in the Lone Star State.

I remember greeting her at the only airport we had, the Houston Hobby Airport. Standing there, waiting for Amina's plane, I looked around with anxiety at the tired-looking airport with its dim lighting and squeaky floors.

What would Amina think? I'd be lucky if she didn't take one look around and instantly change her mind about making the trip.

The memory of her plane slowly pulling up the tarmac is still so distinct. It was a clear summer night, all the passengers had deplaned and my heart was throbbing as I watched stranger after stranger disembark. My anxiety stepped up its frenetic pace just a bit until, finally, I saw her. Amina must have been the next to last person to exit the airplane, but what a sight—my eyes began to tear up with joy.

Emotion swelled within me as I watched Amina cross the tarmac. All she brought with her to Texas was a carry-on bag and a small black trunk. I was unbelievably moved and happy as I counted the minutes until I could put my arms around her, this wonderful woman who believed in me so much she was willing to uproot her life and make that kind of commitment to me.

And so Amina was there with me, as I and hundreds of others at the Manned Spacecraft Center got ready for the big show, the main mission, the one everyone was waiting for: Apollo 11.

News and talk of the Apollo 11 mission and launch consumed the space center and the people who worked there. We didn't have cell phones, e-mails or the Internet in those days but word managed to spread remarkably fast.

Neil Armstrong, Buzz Aldrin and Michael Collins launched from Cape Canaveral on July 16, 1969, with the moon landing set for four days later.

It was like waiting for a Super Bowl game. We all knew it was coming and we all planned for it. We followed news of the various stages of the mission with enthusiasm, knowing a series of sequences was involved. First, the Saturn V, the rocket that fueled the launch, had to perform its task. Then the LM had to go through three stages before going into an orbit of the moon.

Once in orbit, the LM had to travel at 24,500 miles an hour. It took three days for the astronauts to outrun the earth's gravitational pull. Then, the LM entered the gravitational field of the moon itself for two days. It takes a long time and there was a series of milestones that were met along the way.

The lunar landing itself was set for a Sunday, which meant most of the technicians and support staff who worked daily at the space center wouldn't be on hand for the historic moment. But we knew and understood that.

It's so exciting to know that everything we were doing was paving the way for what those astronauts accomplished. I feel that our small contributions were really a significant part of what allowed Apollo 11 to succeed.

And though I was home with Amina on that Sunday, watching the television feed of the lunar landing like hundreds of millions of other people, I remember very distinctly the intimate and direct connection I felt with the mission. I knew all of the switches and lights and dials in that lunar module. I knew the configuration of the cockpit. I had spent so many hours inside of it. I felt like I knew, just a little bit, what those astronauts were experiencing.

But just because we weren't at the space center didn't mean we were unaware of what was going on. Once the Saturn V ignites and clears the tower, Houston takes over and is responsible for that mission, the crew and its success until they are safely back on earth. And once a mission is in progress, the flight director holds the highest position imaginable. No one has the power to overrule the flight director at that point, not even the president of the United States. Once the flight director locks down the room, that's it. There's God, and then there's the flight director.

So when Armstrong and Aldrin began walking across the surface of the moon, and Neil uttered his famous words, "That's one small step for man, one giant leap for mankind," it was all the more exciting and historic for the hundreds of thousands of us who were a part of the team.

My thoughts were with the astronauts themselves, who

risked their lives inside these vehicles in the harshest environment man has ever known. Those brave men were truly
courageous, relying on the quality and construction of the
space vehicles they rode in to keep them safe.

And I had another connection: the plaque that went to the
moon, left there by Armstrong and Aldrin. The plaque that
read: "Here men from the planet Earth first set foot on the
Moon, July 20, A.D. We came in peace for all mankind." I had
an opportunity to hold the original plaque in my hand prior
to it being placed on the vehicle. I'm sure the white gloves I
was wearing prevented my DNA from making the trip, but one
never knows.

It was like a party in Houston after the moon landing, and
post-Apollo 11 seemed the ideal time for Amina and I to do
what we had been longing to do: get married. Since Houston
wasn't all that far from Mexico, I thought it would be different
and kind of exotic to get married down there.

What was I thinking? We went and got our blood tests,
filled the car with gas and drove off toward the border, but
it wasn't until later that we realized we couldn't get married
there.

So we stayed on the Texas side of the border and drove
to the very first justice of the peace we could find. We were
in Laredo, Texas, and I can remember sitting on the court-
house steps waiting for the office to open. I'm sure our parents
wouldn't have been thrilled if they had realized this is where
we'd end up. But they knew we were serious about each other
and I guess they realized there wasn't much they were going to
be able to do to stop us.

So it was on July 30, 1969, that Amina and I were married
by the justice of the peace. He had to get the clerk from across
the hall so that we could have a witness. It must have taken all
of seven minutes.

Then the woman in the office there handed me a white plastic bag with several samples of small bars of soap, the kind you see in hotel rooms, and several boxes of Tide detergent. She gave us a big smile and wished us well.

There were no gifts, no friends, no family present to wish us well. There was no lavish and expensive party. It was just Amina and I, and it was fine. We were just so happy to be married.

At that point, we felt we had a choice to make: go home to New York or stay in Houston. I had become quite fond of this southern city, and felt there was a lot of potential here. We talked it over for a while, weighing the various possibilities and what-ifs, before we decided to take a chance. That meant forgoing a return to the Big Apple and putting down roots in Texas.

I tried to hold onto my job in Grumman for as long as possible. The company's generous treatment of employees like me, who had volunteered to relocate to Houston, helped us greatly. After all, I had worked countless hours of overtime, and the company had been picking up the tab on the rent for my one-bedroom apartment the whole time. It had given us the chance to become "thousandaires," so to speak.

Amina and I were feeling good and thought we had enough for a down payment on a house. We found our home, 938 Redway in Clear Lake City, a suburb of Houston nearby the Manned Spacecraft Center. The house cost us about $24,000; I found out later the previous owner had built the small, ranch-style home for just $16,000, and thus made a profit by selling to us. But at that point, with a thirty-year mortgage and a $150-monthly payment, it didn't really matter. We were committed.

Amina went to work for NASA in an administrative position, which was a big help to our household finances, especially when things at Grumman started to get a little tense.

With the space program winding down, it was only a matter of time before the company's presence in Houston began to shrink. Grumman had moved the remaining technicians off the NASA site and set us up across the street in another facility. Things were drastically different now than they were just a couple of years before at the height of the space program. Technicians were now doing whatever they had to in order to keep their jobs in the face of radically different job assignments and a lessening of overtime.

Some in management did what they could to try to sustain the community of Grumman workers in Houston, trying to secure additional work for us from company headquarters in Bethpage, New York. There was some potential, since Grumman technology was also being used in the medical industry and Houston had become a center of the medical world.

It's amazing what you will do to keep your job and maintain your dignity at the same time. And as time had gone on, I felt an intimate connection to the mission of the space program—a connection that outweighed any desire to move on to whatever Grumman might have had in store for technicians like me. Eventually things settled down a bit, and we all went about our lives hoping Grumman would be able to get more work from NASA, or that, somehow, the people in Bethpage would forget about those of us left in Houston.

The uncertainty didn't dampen any of the plans Amina and I were making. We talked a lot about starting a family, having a couple of children of our own, maybe adopting another along the way. It was something we were pretty serious about and we were thrilled to find out in 1970 that Amina was pregnant.

With Amina and me establishing our life together, I was full of feeling. I was excited and scared, but, most of all, I felt reflective. Becoming a father made me think about my own childhood and the way my siblings and I grew up.

I knew then that without the social safety net provided by welfare, my family might have met a different fate. I was grateful for that and, most of all, I felt like I wanted to give back.

I also knew that even though our lives turned out to be pretty good, and that my family loved and took care of me as best they could, it would have made a huge difference to have someone around to talk to, someone who could give advice, encouragement or just a friendly ear.

I don't remember how I chose the Harris County Youth Village, a facility for troubled boys, the kind of kids who weren't bad enough to go to jail, but who were in danger of ending up there if they didn't clean up their act. Maybe I drove by the building, which was pretty dilapidated, or someone told me about it.

But it was a convenient location and, for me, a special mission. I felt like I had the time to give and the desire to help. My mother had always instilled in me how important it was to do the right thing and to just be honest.

I was honest with myself and thought: *You know, you can do this. You can start out a little bit at a time.*

After all, sometimes no matter how hard parents try, for whatever reason kids can be more willing to listen to a piece of wisdom or guidance from a mentor instead.

So I started volunteering. The Harris County Youth Facility was always looking for volunteers and people to help spend time with the boys, and I felt like I could relate more to them.

I had never done anything like this before, but I found myself talking to the kids, spending time with them and sometimes taking them out for the day. It was comforting and reassuring to me that I could reach them.

Sometimes you see yourself in a child, and I saw myself in some of those troubled kids—that feeling like you just need someone to listen. I knew that feeling well.

Kids are like clay; if they get the proper direction, they can become anything in the world. And it's so easy for a child, a teenager, to make a wrong turn. I've come to some forks along the way myself, and I truly believe if it weren't for some sort of divine hand helping me to make the right choices, my life might have become something quite different.

Maybe in a way I was practicing for the day my own son would be born. But the more time I spent helping kids, the more my heart grew and the greater my reward.

Time passed, and as the pregnancy moved along, Amina and I were both working. One day in March of 1971, she had left work to go to the doctor. During the visit, the physician told Amina it was time for her to go to the hospital.

So did Amina give me a phone call? Let me know what was going on? Nope. She got into her car and started driving herself to the hospital.

While she was heading there, a police officer stopped her car and she had to explain that the doctor had ordered her to go to the hospital. When she got there, she checked herself in, got settled—and then called me.

Here I was, about to become a father for the first time, and I didn't even get the privilege of driving my wife to the hospital even though I had practiced doing so. I only found out she drove herself after the fact.

When I finally did find out what was going on, I left work and immediately went to the hospital to be by Amina's side. She was on pain medication, but as the hours dragged on, it was clear this was going to be quite an ordeal.

Amina was in labor for twelve hours. She was in and out of awareness and in great pain. I was at my wit's end, watching her struggling and in agony. There wasn't much I could do other than hold her hand and try to comfort her.

The frustration and helplessness I felt watching my wife

suffer with this immense physical exertion were relieved only by the birth of our son, Kumar David Cisco. The joy of his arrival, coupled with the flood of gratitude that Amina's physical pain was over, brought my emotions to the surface.

I remember that, when all was said and done, I had a moment of solitude and was struck by the weight of responsibility brought on by my new title: Dad. I looked at myself in the mirror, feeling more love and fear and hope than I ever had before, and the intensity of my feelings brought tears streaming down my face. I thought: *I've got to be a good father. I've got to be a good husband. I have to provide for the family and be a role model, and I will.*

Soon we were able to bring Kumar home and begin this newest phase of our lives together, that of being parents. But more changes were afoot as things at Grumman became increasingly unstable for the company's workers in Houston.

Though things were not all bad. In April 1972, I went down to see the launch of Apollo 16, which was quite exciting. Commander John Young, CSM Pilot Ken Mattingly and LM Pilot Charles Duke were the crew for the Apollo program's tenth manned mission.

Just two years earlier, Mattingly was supposed to take part in Apollo 13, but was pulled from the crew with just a few days to go before the launch. At the time, it was Charlie Duke who somehow managed to expose Mattingly to the German measles. The possibility that he would get sick was enough to keep Mattingly off the flight.

It was fun to go see the mission launch, but it didn't help the overall picture at Grumman's Houston operation in those days. As time went on, the work started to run out and the company started to lay people off. I had attained a good position by then, but no one was safe from the ax.

A workplace becomes poisonous enough when some are

being let go; it's worse when you are the one involved, in any way. My position afforded me no real authority or power, but when it was time to tell employees they were losing their jobs, I was appointed as some sort of grim reaper. I was given the names of those whose jobs were on the chopping blocks, and tasked with sending them in to see the bosses. I was never asked for any input.

It was the lowest point of my career at Grumman. I could barely stand it. When all the layoffs were complete, the only relief was that it would probably be a few more months before any of us had to face that horrible situation again.

With the atmosphere at work becoming more uncertain and difficult to deal with and a young family to take care of, I knew I had to do something. It was time to make a change, so I decided to get out of the technical end of the space industry and pursue a different direction.

It looked like a California-based company called Rockwell International had a field office that was going to become the primary contractor for the space shuttle program, so I applied for a job with them and was hired.

PART III

I left Grumman after nearly a decade with the company, and began working for Rockwell International on March 5, 1973. By now, Amina was pregnant again, and I was set on moving toward a more secure position in my career.

This move was supposed to plant me squarely in an administrative position as opposed to a technical one. No more flipping switches for me.

I welcomed the change but it didn't last long. After getting hired, I was quickly called upon for technical assistance with a critical project: the United States Skylab program.

Skylab was the first and only space station NASA would launch on its own. The goal was to create a facility in space that would give astronauts a place to conduct experiments on everything from how humans adapted to space, to studying the electromagnetic spectrum of the sun and observing the earth. It was the result of a NASA program that began in 1965 with the goal of adapting Apollo vehicles and spacecraft for other purposes.

On May 13, 1973, NASA used a Saturn V rocket to launch Skylab into space. But just minutes after takeoff a key component was torn off. Without its thermal shield, Skylab was exposed to immense temperatures and became extremely overheated—to the tune of 140 degrees.

This problem put the whole Skylab program in jeopardy. There were three crews scheduled to visit Skylab, but it was too dangerous, the heat too much, for any human to stand it. NASA needed a solution, and a team of people was assembled to tackle the problem.

I was asked to be a part of that team, and found myself once again plunged into the hands-on, hardcore work of making sure our astronauts would be safe as they ventured out into space.

The heat was on us, you could say, to come up with an answer that would make Skylab habitable. Eventually, this team created a thermal protection parasol. We worked tirelessly, diligently giving up our nights and weekends to develop the heat shield. This device wasn't pretty, but it created a barrier between the Skylab and the dangerous solar energy that threatened to cook anyone who tried to set foot in the space station.

Now it was safe for all three crews to go up and take their respective turns manning Skylab. Normal mission activities were resumed for the entire program and the mission met or exceeded expectations. Crews went aboard Skylab for twenty-eight-, fifty-nine- and eighty-four-day missions. The last mission set an endurance record for space travel at the time.

I was so proud to be a part of this team that was able to keep America's space station in the sky. It became a huge step forward for NASA as the space program was trying to transition away from lunar landings and on to other projects.

This hectic time dominated my opening weeks and months at Rockwell. Before I knew it, it was early October, 1973, and time for Amina to deliver our second child.

I was in the middle of a long, hard day at work and didn't know it, but again she drove herself to the hospital before calling me. When she did give me a ring, I told her I'd be there as soon as I could.

When I arrived at the hospital, Amina was sitting up in the bed, chatting and calm. Unlike when she was in labor with our son, Kumar, this time Amina was going to have a natural birth. We didn't know exactly what to expect, but I prepared myself for another long, tense experience of watching my wife in a painful delivery.

I had already worked a ten-hour day, and I hadn't eaten. Everything was fine at the hospital, and Amina was quite relaxed.

"Why don't you go grab something?" she asked. "It's going to be a long night."

"No, no," I said.

But she pushed and finally I said okay and told her I would be right back. I thought to myself, *Really, what could happen?*

I got in my car and drove to a Kentucky Fried Chicken, right on the corner near the hospital. I went in and sat down to eat; it couldn't have taken me any more than thirty minutes to finish, get back in the car and make my way back to my pregnant wife.

Returning to the hospital, I spotted a nurse and stopped to ask, "How is she?"

I was expecting there to be pain and long, grueling hours of labor. I was expecting another uncomfortable, disturbing time sitting on the sidelines, watching my wife suffer. So when the nurse said, "You need to go up there," I didn't know what to think.

I rushed up to Amina's room and found her there, sitting up in the bed just the same as when I left her to go eat.

"Oh, my God, this is going to be a long night," I said.

"No," Amina replied. "I already had her. She's down the hall."

I stood there for a moment, stunned. My newborn daughter,

Lahna Mia Cisco, had been born, and I missed it. I was eating fried chicken while my wife brought our beautiful baby girl into the world in less than forty-five minutes.

How, I thought, *do I live this one down?*

At least it was good chicken. But to this day, whenever my family and I go past a Kentucky Fried Chicken, everyone looks at me.

So here we were: husband and wife—a two-career couple with our own home, a son and now a daughter. Life was good, and one of the highlights of my career was just around the corner as those who worked on the Skylab heatshield were given one of NASA's highest honors.

Our work on the heatshield brought many recognitions and awards. I received a photo of the Skylab launch, personalized and autographed by the three crews of astronauts that flew on the space station.

But the most precious honor was one that still makes me proud today. NASA's Silver Snoopy award is one of those accolades that stay with you for a lifetime.

Less than one percent of NASA workers and contractors are even eligible to receive it each year, and it is only given to those who have significantly contributed toward the success of a mission or helped achieve improvements in design, techniques, safety, etc.

This isn't an award usually given to management, or in recognition of those who have put in many years of service. This is an award meant to recognize hard work, good work that makes a difference. It is an award handed out only by the astronauts themselves, and almost always in front of a recipient's coworkers.

It is a once-in-a-lifetime event to get a Silver Snoopy. I got mine on April 15, 1974. I was quite moved when astronaut

Alan Bean, the fourth man to walk on the moon, presented me with my sterling silver lapel pin, which is in the shape of Snoopy himself dressed in a space suit, a design based on a drawing by Peanuts cartoonist Charles Shultz.

Each award comes with a letter of commendation and a certificate that reads: "For professionalism, dedication and outstanding support that greatly enhanced space flight safety and mission success."

The Space Flight Awareness program hands out the Silver Snoopy and other NASA awards. Since 1968, only around 12,000 people have been awarded the Silver Snoopy.

The Silver Snoopy was a career highlight, one I will never forget.

For a time, I never thought I would walk away from spacecrafts and astronauts, NASA and the Johnson Space Center. After Skylab, I was also briefly involved in the Apollo-Soyuz Test Project.

The ASTP marked a great departure from the origins of America's space program, which was born out of a competition with the Russians in what would become a political and technological race to see who would be first to land a man on the moon. In a boost to national pride, America won that race.

But less than a decade later, things had changed and, now, NASA was working with the Soviet Union. The July 1975 mission was the final space flight of the Apollo program.

I was so proud to be involved in the ASTP, though my role was mostly administrative and not nearly as hands-on as it had been during the development of the thermal shield for the Skylab.

Eventually the move I began to make when I left Grumman and joined Rockwell International took hold, and as the work with Skylab came to an end, I settled into my new role as an administrator.

My boss was a man named Tom Short. He was a vice president of field operations at Rockwell, and he became a mentor and lifelong friend of mine.

I spent most of my time at Rockwell as the affirmative action officer. This was an entirely new endeavor to me, and I was challenged both by the work and by my own insecurity.

I was the affirmative action officer and oversaw other personnel functions for the company, which meant finding the right people to work for Rockwell, but I was responsible for conducting background checks on the candidates and then processing them through company security. The challenge put before me was to bring in new, young, diverse talent.

The work was new to me but as I started to earn my stripes I moved up the ladder, and was promoted to a small office just outside of the vice president's. Now to keep it in perspective, my office was the equivalent of an oversized closet; however, it was my very own office.

Tom Short was a person who always greeted you with open arms, had an affectionate, infectious laugh, a belly laugh, and was one of those people to whom everyone could relate—even if he was the boss. I never met anyone who knew Tom Short and didn't like him. He was always working to make sure everyone was a part of the process, to make sure everyone felt like part of the team.

Even though Rockwell was a progressive and forward-thinking place, I often felt uneasy in those first few months. I was fortunate to have someone who believed in equality, and presented opportunities; it was up to you to qualify and to face the challenges.

I felt there were several people there who were really pulling for me to succeed. Herb Harman, a manager, was always

This NASA photo shows JFK declaring we are to "become the world's leading space-faring nation."

Posing at the same podium where JFK declared, "We choose to go to the moon."

Gene Cernan, fourth from right; far right is Tom Short and me.

"WE CAME IN PEACE AND WITH HOPE FOR ALL MANKIND"....
Apollo XVII Dec 1972

TO DAVID - MY SINCERE THANKS FOR ALL YOUR EFFORTS
IN HELPING MAKE THIS MEMORABLE MOMENT POSSIBLE -
GOOD LUCK & BEST WISHES.
Gene Cernan 10/22/96

Apollo 17 Commander Gene Cernan, last man on the moon.

MANNED FLIGHT AWARENESS

APOLLO 8

IN APPRECIATION FOR YOUR CONTRIBUTION TO THE APOLLO SATURN PROJECT. THE APOLLO 8 CREW CARRIED METAL IN THIS MEDALLION ON MAN'S FIRST FLIGHT TO THE MOON

Although many thousands of us have worked to make our manned space flight possible, only a few of us can experience the incomparable travel in space.

It seemed to us that there should be some way of providing those who worked in the program some symbol of their efforts. We carried on the first manned lunar flight a small piece of aluminum. This piece of metal was melted down and incorporated into the metal used in making this commemorative medallion.

I hope that this will serve not only to give you a sense of active participation in this historic flight, but also to assure each of you of the appreciation of Astronauts Borman, Lovell and Anders for the superior workmanship, conscientious performance, and dedication to the safety of fellow human beings that made our successful flight a reality and will do the same for the missions that remain ahead of us.

Frank Borman

At the Kings Inn, Dave Cisco was on hand to shake hands with the Russian cosmonauts and U.S. spaceman Frank Borman.

Apollo 8 award and autographs from Frank Borman and a Russian cosmonaut.

A medallion earned for my work on Apollo 11.

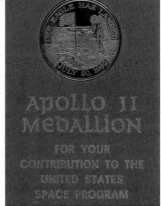

Last voyage to the moon, signed by the Apollo 17 crew.

Alan Bean, Apollo 12 astronaut, lunar module pilot and fourth man to walk on the moon, presenting the astronaut's highest honor—the Silver Snoopy lapel pin.

Admiring my father's Army photo.

My father's Army photo.

My mother and father, 1940s.

My wife, Amina, and me.

My son, Kumar, and his mother.

Lahna, and dad.

My grandson, Xavier, my son, Kumar and me.

Me and my grandson, Hayes.

My granddaughter, Anika.

With Mother at city hall, standing before the bronze plaque
noting me as city councilman.

Mother and me on one of our special getaways.

My buddy Joe Cucci on the job, and ready for action.

Captain Buddy Howard and me.

Six female first officers hired at Texas International Airline.

NATIONAL AERONAUTICS AND SPACE ADMINISTRATION
LYNDON B. JOHNSON SPACE CENTER
HOUSTON, TEXAS 77058

REPLY TO
ATTN OF:

Mr. David L. Cisco
938 Redway Lane
Houston, Texas 77058

Dear Dave:

We of the Astronauts' Office take special pleasure in commending
you for the dedication you have demonstrated in the accomplishment
of your assignments on the Skylab Program.

The problems encountered during the launch of SL-1 resulted in a
crash program involving around-the-clock activity in order to build
hardware, make changes to the CSM and provide stowage for the new
hardware and tools to repair the workshop. After the successful
repair of the workshop by the SL-2 crew, normal mission activities
were resumed for the entire program and the meeting and exceeding
of mission objectives. This required long hours and constant
attention over a period of months, including the Thanksgiving,
Christmas and New Year holidays.

Your personal expertise and dedication in assisting in the fabrication
of the Thermal Protection "Parasol" which resulted in making the workshop habitable is recognized by this presentation.

On behalf of all the astronauts, it is my pleasure to present to you
the Astronauts' "Silver Snoopy" award for professional excellence.

Sincerely,

NASA Astronaut

Commendation letter from the NASA Astronauts' Office.

My visit to the White House.

Red room at the White House.

A napkin snagged from the White House reception.

Standing in front of the famous JFK painting in the White House.

Thomas Jefferson's presidential chair at the White House.

Monitored STS 116 Shuttle Sim-Run Mission Control, 10/24/06.

Celebrating my successful landing in the space shuttle simulator.

I took this photo of Neil Armstrong and Buzz Aldrin at the Apollo 11 fourtieth anniversary celebration.

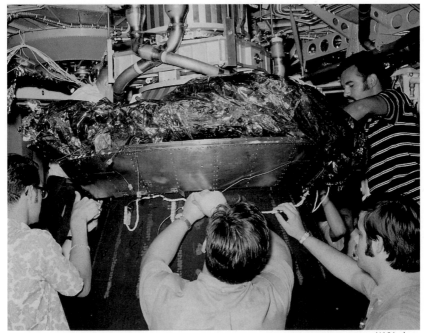

NASA photo

Preparing the LM2 for its trip to the Smithsonian in Washington, DC. To my immediate left is Richard Specion.

Two Mylar swatches presented to me were from the thermal coating of the command modules of Apollo 11 and 13.

A plaque celebrating the moon landing and containing metal from Apollo 11.

Tom Savage, police chief of Taylor Lake Village, presenting me with an award.

To Dave,
With sincere appreciation for your support of the Skylab Missions. Our warmest personal best wishes —

NASA-S-73-26911

Sky Lab launch, signed by all nine crew members.

To Dave Cisco —
    Here's hoping the world will always look as good to you
as it looked to me when this photo was taken! It's been
great to work with you. Best wishes for success and happiness for both you and your family in your new venture.
Second spacewalk - Skylab II - 28 July - 25 Sept 1973
                                    Jack Lousma

Astronaut Jack Lousma, second space walk, Skylab 2.

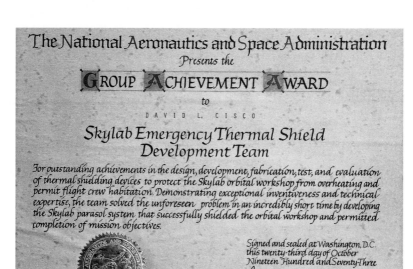

Group achievement award, Skylab.

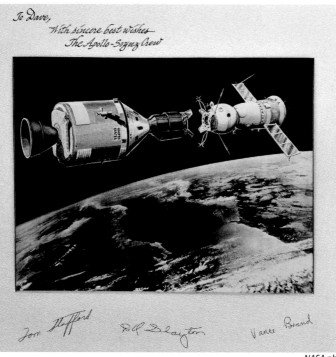

NASA photo

Joint US and Russian ASTP, signed by the US crew.

offering me encouragement. He was a talented man who seemed to like everyone.

Since I would be hiring new and fresh talent, one of the people I worked with frequently was Ben Boykin, the director of engineering. He was an active guy. He would sit you down at his desk, and ask you what you thought about a given situation. He would listen carefully. Then he would lean back in his chair, and, before you knew it, would go in the opposite direction, taking the time to methodically go through and explain his reasoning.

Once his mind was made up, he seldom changed it. I don't think he cared if he bruised your ego or not. What concerned him was whether he was making the best decision. He had a way of making it sound so final.

His way of doing business reminded me of how people in Houston often reacted when they found out I was from New York. "You know, people from New York... they're rude," is what people would usually say to me.

Then I would correct them, and point out that, though that may be the case sometimes, people in New York are mostly just direct and eliminate the long way around. Ben wasn't from New York, and he wasn't rude, but he had that directness about him. He was one of the people at Rockwell I always admired for cutting through the smoke and getting to the other side. Ben taught me that things are not always going to go your way, so pick yourself up, dust off and continue in a positive way. If he could have bottled his style of management and sold it—wow!

The job gave me a certain level of security. It also gave me more visibility. After all, I was moving up in the world.

I still had my old 1968 Cutlass Oldsmobile in which I had driven from New York to Houston. It was a little more than five years old then.

Another engineering manager at Rockwell, Mort Silver, was the kind of guy who liked to buy used cars at a discount. Mort was a very intelligent man with strong opinions. He'd buy these cars, take his time with them and work on refurbishing them. Then he would sell them for a profit.

I was looking around for a sports car, and Mort mentioned that he had one he was thinking of parting with. Maybe the car needed more work than he had time for.

It was a 1968 Mercedes 250 SL Roadster. It was sleek. This car was the picture of cool, and had a removable hardtop.

The only problem with it that I could see was that every time you started it up, it was blowing a puff of blue smoke out of the tailpipe. When I heard that, my technical senses kicked in, and I figured it needed a new piston ring or a valve job.

Sensing a good opportunity, I haggled with Mort for a while, going back and forth with him over the price before we finally came to an agreement. I handed him some money, and signed over my Oldsmobile to complete the deal.

I really hadn't done any research on this car I was buying. And it wasn't like nowadays, where you can plug a few words into Google and get reams of information back in a few minutes.

It didn't really matter. I knew I had always had good karma and that things like this usually seemed to pan out. Besides, this was a cool, classy car—even if it did look like I was getting the short end of the stick.

Of course, that wasn't really enough to sell Amina on what I had done. After all, I had sold my car only to get one of the same year in return, one that needed work. But in the end she didn't object too much.

Over the years I have rebuilt the engine, refurbished the interior to its original condition and have put a lot of time and work into the car. But I never really knew what I had until recently, when I was ordering some parts for it from a car dealership in California that specializes in these classic Mercedes.

When I told the man the VIN, year and make of the car, I was greeted with silence on the other end of the phone. He asked me to repeat myself and I did.

"I guess you know what you have?" the man asked slowly.

"Yep, I know," I said.

No. No, I didn't know. I didn't know at all. It was then that I found out just what the man was talking about.

My little Mercedes—the one I used to drive every day back and forth to work—was extremely rare. This particular car was one of only a few thousand that were produced between November 1966 and January 1968. In fact, in 1968, there were just two such Mercedes even made. The only question was which one of the two did I have?

In 1968, it was selling for a grand total of $6,897. I know now it's probably worth a whole lot more than that. So, these days, I'm extra careful when I drive my little sports car, which only gets taken out on nice, clear days when there are no clouds in the sky. I'll take her out for forty-five minutes or so at a time, and then back in the garage she goes. After all, my ride is 100 percent restored and is more than forty years old.

Once I found out its true history, I got an estimate that put its worth at possibly six digits.

■ ■ ■

Each Friday afternoon, Rockwell International would have staff meetings and the members were usually senior management or director level. I was neither. But Tom surprised me by always making sure that I was a part of those meetings and that I had a seat at the table.

For the first few months, I was uneasy with the group. But

after a while, it became more natural, and I became more comfortable. I stopped noticing the differences between me and the other people at the table, and I started to feel normal being there and being a part of that group. I started to find my own voice and contribute my ideas.

Those meetings, that experience, helped to open windows in my mind. I had always had a generally positive, can-do attitude. There wasn't much that I ever let hold me back.

But at the same time I wasn't ignorant of who I was, or where I came from. I wasn't the type of guy who was ever meant to be sitting at that senior management table. At least, I'm sure that's what some people must have felt. But for Tom Short to take that chance on me, and give me the opportunity to discover a different kind of confidence in myself, was one of the first times anyone had done that for me.

Once you're given a chance like that, you become one of the boys and it gives you such a degree of confidence that you can't help but move to the next level.

Looking back at that time, I realize it was the beginning of a turning point in my life. Growing up, we didn't talk a lot about the future. There were no grand expectations or concerns about social status. In some ways, that was good. But it also meant that kids like me were barely given a second look, lacked the encouragement we probably needed and were most likely denied resources.

It's funny. As a society we can eradicate diseases, create impressive technology or even put a man on the moon. What about children who are struggling with hunger, or who don't have clean clothes? What about kids whose parents work so much just to make ends meet that they can't be around to parent? What about the young ones who are learning that success is about money, fame and living on easy street, instead of doing what's right, working hard and finding your own path?

The secrets of success are really quite simple, if not all that glamorous. A person decides their own definition of success—for me, it meant financial independence, being able to take care of my wife, my kids, my mom and dad, and giving back to my community.

But the opportunity has to be there. And, at some point, when a person comes across or makes their own opportunity, they have to apply themselves.

You don't have to be the valedictorian. And you don't have to be the company president, movie star or the professional athlete who gets the big money contract.

You don't have to be the astronaut who lands on the moon to be successful. It's not the only way. There are thousands of ways a person can achieve their dreams, a hundred different paths to success. There's truth in the cliché: "Reach for the moon, and you'll land among the stars."

In some ways, I'm sure, I had already exceeded a lot of people's expectations of me. If that high school guidance counselor, the one who told me to go get a factory job, could have only seen me at Rockwell!

Tom Short did much more for me than to give me a seat at the grown-up table, so to speak. He also gave me one of the most devastating, and important, moments of my life.

How, or why, we began this particular conversation I couldn't say. Tom was such a nice, generally positive guy that when he made his statements I didn't know what to think at first.

"You know, David, you're probably not going to go any higher at this facility," he said in an even, monotone voice that betrayed nothing. "The only way you're going to move up the ladder is to go to Downing, California, and go to the Rockwell corporate office."

My heart broke just a little bit when he said that. I had

felt like I was on top of the world at Rockwell—working in the administration, having my voice heard and being an active part of the company.

When he mentioned California, I found myself flashing back to my Grumman days, and remembered how it felt when the work was drying up in Houston, and how much I hadn't wanted to return to Bethpage. There I was, a New Yorker who didn't want to get sent back to the East Coast, or to the West Coast. All I wanted to do was stay in Texas.

But Tom didn't leave me hanging. He kept talking. He pointed out that he himself was a vice president, and it was unlikely he would ever go higher than that.

"David, you're a smart guy. You should go into your own business," he said. "That's what I'm going to do."

It took a few days for his words of wisdom to settle in, and for me to really hear what he was trying to say. Go into business? I didn't have the first clue what that actually meant. It was like standing on the edge of a swimming pool with my foot hovering above the water, and his encouragement was just like having someone tell me to go ahead and stick my toes in.

Time passed, but his words stuck. It started to make sense—after all, once I made my mind up, I could do anything.

PART IV

The more I tossed around the idea of going into business, the more intrigued I became, but Amina wasn't exactly excited about it. I couldn't wait to talk it over with her but when I told her about my discussion with Tom, her reaction, I'm sure, echoed that of many wives: "Go into business? Are you crazy? What kind of business? How would we pay for it?"

I didn't have much of a comeback at that point, so I tabled the discussion. Amina had made some great points, so I quietly started to do some research. If I did this, what kind of business would it be? Who would be my customer base? What would I be doing?

I ended up looking into just about every small business there was. Not a very scientific way of doing things, but it worked for me. I kept coming back time and again to the travel business. After all, there were a lot of big companies in Houston, and they sent a lot of people back and forth all the time to their various corporate offices. So why not?

I looked into what it would take to qualify to become a travel agent and get the proper licenses, bonding, and training. That took some time, but once I felt like I had enough ammunition, I brought the topic back up to Amina.

I'll always remember that night: She had worked all day at NASA, had Kumar, then seven-years-old, and Lahna, then around five, to take care of, and was making dinner.

As Amina was stirring her special meat sauce that would go over our pasta, I gingerly mentioned that I had "solved the problem" of the idea of going into business for ourselves. I told her I had narrowed it down to the idea of a travel agency.

A beat passed as she stirred, but Amina suddenly stopped and looked at me.

"Okay. Tell me some more," she said, and then wasted no time in asking how we could do this on our current salaries—not to mention with two kids.

Well, I hadn't exactly done all the math, so some of my answers were a bit sketchy. But I assured her that I would be able to get the answers.

That began many months of research, of discussions each night after dinner, of planning and weighing the pros and cons. We talked about it for so long that anything seemed possible. Eventually, we put ourselves on the path of such a big change with some baby steps, and Amina signed up for a course given by the American Society of Travel Agents.

Actually, Amina was the one who continued to forge a path for us into the travel agency business while I continued working at Rockwell International. After finishing her first ASTA course, along with other training, she decided to get some hands-on experience—a major step forward toward making this dream a reality.

So Amina took the leap of faith first, and quit her administrative job at NASA to work for a travel agent. She got a job with a local agency called Kay's Travel, which later became TV Travel. This was the largest travel agency in the Bay area, and just about all of the companies subcontracting with NASA used it. My wife is not only beautiful but also very smart, and once she puts her mind to something, she gets it done.

Things at Rockwell were slowing down a bit. When I got

the job there, I discovered dear friends who gave me wonderful opportunities to find my own potential. Now, toward the end of 1974, my boss and close mentor, Tom Short, was leaving the company. Tom was gearing up to do just what he had said he was going to do—he was going into business for himself, and would become the CEO of an oil business.

Seeing Tom go was sad. He was the most compassionate, motivating person I had ever come across and, now, he wouldn't be there for me. But it was also an inspiration. After all, this was the guy who helped me see possibilities I hadn't dreamed of and here he was, following his own advice. It was encouraging.

I stayed at Rockwell until 1976, when a corporate headhunter approached me. I wasn't necessarily looking for another job, what with Amina and me solidifying our plans for our own business. I didn't know at the time that this person was trying to recruit me for another company. I must have said something to someone, though now I couldn't for the life of me remember what. What was I thinking!

Well, I certainly had nothing to lose, so I listened to what the recruiter had to say. He set up an interview for me with Texas International Airline. This was the company that would eventually take over Eastern Airlines and Continental Airlines. The job opening the recruiter was trying to fill was the manager of employment system-wide.

This was no small thing. Every job in the company would be going through my office, if I got the job. All promotions, everything from the airplane mechanics to the cabin service crews to the pilots and company management, would be the responsibility of my office. This was a big job.

It's amazing how much one's frame of mind makes a difference when you're interviewing for a job, especially if you're not really sure you want the job. You seem to do better. I had agreed to go in for an interview, but, psychologically, I was in a different frame of mind than your typical eager job seeker.

I had a full day of interviews with senior management—including Sam Coats, the corporate attorney; Don Breeding, the vice president of flight operations; Ed Cathel, the vice president of personnel; Bob McAdoo, vice president and controller; and Jim Arpie, the vice president of maintenance.

The heavy-duty interviews showed how big a deal this was. This was a key position for the airline and they wanted to make sure they had the right person. At that time, Texas International had aggressive plans for growth and they needed to have the right person to help staff positions at all levels and sectors of the company.

It was an exhausting process to go through all of those interviews. I kept thinking: "What have I gotten myself into?" However, they wanted me to come in again for a second round of interviews, so I must have done something right.

This time, they brought out the big enchiladas—including the chief executive officer, Frank Lorenzo. But I felt good after this second round, which included meeting with Don Burr, the senior vice president, and Rob Sneaker, the treasurer; I seemed to have good chemistry with all of them. I may not have been looking for a change like this, but I knew I could work well with these people and this company.

When it was finally time to meet with Frank Lorenzo, I was feeling pretty good, based on how the other executives were treating me. I met with Frank in his executive, wood-paneled office. We sat on the sofa and talked for about a half an hour—that's when he told me this was just a formality because his staff had already decided to overrule whatever he said.

I always felt Frank Lorenzo was a perfect gentleman and

a wonderful corporate officer who had a lot of great ideas. He was ahead of his time. After all, he was the one who mentioned to me, in the late 1970s, that deregulation of the industry was coming and that we needed to prepare for it. Ultimately, he predicted, there would probably end up being four or five major airlines and that's it. He was right on the money; that's exactly what's happened to the industry in recent years.

I was officially offered the job on September 20, 1976, and it may sound strange, but in the back of my mind, I felt like this was just a temporary thing. It would be helpful to get me to the goal Amina and I had set for ourselves: running a travel agency. This was a related field, in a way. This was a transition job, I reasoned, something to help get me closer to being my own boss.

Oh, what a "transition job" it turned out to be! Texas International was a big company, one that was infested with the good ol' boy mentality. There was very little diversity in the staff at that time, particularly in key positions like pilots. To the company's credit, I was given the challenge of changing that culture, and was to play a key role in building a team that would reflect all of the people we served.

Just because Texas International was striving to diversify its workforce didn't mean that the old mindset wasn't alive and well. Heading the company's personnel office also meant that individuals from various locations within the company had no problem sending family friends in to get easy jobs. Those people weren't very happy after I came on board.

There was a lot of resentment at first since they couldn't understand the concept of posting open positions in-house, giving current employees a first crack at the opportunity, and then interviewing and finding the best candidate for the job. Doing so meant opening up the process of how people were hired, and opening up job opportunities in a way that considered seniority and change, such as the possibility of female pilots and male flight attendants. Those who met the qualifications of the positions for which they were being considered were included and considered fairly for each open job, and I committed myself to that end.

This was a great young airline; the top management was made up of people whose average age was just forty-two years old. They now stood behind me and encouraged me to go down a path that reflected the best ideals. Texas International was growing so fast that the pressure was really on to find the right staff and to fill these jobs that were being created. The lack of staffing was felt throughout the company, and some airport station managers had to deal with delays at their stations because there simply weren't enough people to go around.

But I had been given a mission with a specific set of rules— to find the best, and only the best, people, to embrace diversity by giving people of all backgrounds a chance, and to give our own employees an opportunity to move up the ladder.

The process took time because we posted each job in-house first, but I had other staffers working as hard and as fast as they could to fill these slots. One day, one of those staff members asked me to interview a candidate for an entry-level position since he was going to be out of the office that day. I thought: *Well, how long could it take?* and agreed to do the interview.

I was pretty busy with backlogged work myself. But this was an entry-level cabin service job and, again, I figured it really couldn't take more than fifteen minutes. I admit that when the man walked into my office, my mind was someplace else, drifting to the piles of work I still had to get done.

I could hear the man talking to me as we went through the interview. He seemed nice, was neatly dressed and quite well spoken. At five-foot-ten, with his weight in proportion to his height and a head of sandy blond hair, this man seemed quite humble.

Distracted, I am not sure how much I was listening. I began to okay him for processing when I stopped myself. This wasn't the way we did things at Texas International. The process that I had put in place called for a lot more thought and attention than this, and that's something any job candidates deserve, even if they are only applying for an entry-level job. We still had to make sure we were hiring the right person.

So I asked him to repeat a few of the things he had mentioned earlier and that's when it started to dawn on me that some of the things this man was saying just didn't jibe. I've interviewed hundreds of job candidates and can tell when something just doesn't fit.

I got right to the point and told him I had a problem with what he was telling me. I needed more information. By then we were thirty minutes into what I thought was going to be a fairly quick interview.

That's when he looked me square in the face and said: "Mr. Cisco, you're right."

He swallowed and paused for a bit before he went on.

"I used to work for the phone company, and I listened in on a conversation," he said slowly.

Right away, I felt that wasn't good. Then the man proceeded to tell me how he had called the person in the conversation back, and was eventually reported—and fired.

At that moment, as he made his confession, he realized the interview was over. But he sat there, with tears streaming down his face, and kept talking. It was as if he just needed to tell someone. The incident had actually happened several years before. He had apologized but things still snowballed until he lost not only his job, but also his wife and family. This was a man at the end of his rope.

For some reason, I don't know why, but this verse came to mind: "He who is without sin, cast the first stone." This was a phrase I often thought of, and it came to me as I watched this weeping, broken man. This was a person who was just trying to get his dignity back. I took a moment and thought about this, and made the decision that I would hire him.

Some might have seen hiring that person as a risky move, but it was a chance I was willing to take. That man was so thankful. He made a personal commitment to me that I would not be sorry for taking a chance on him. He thanked me for trusting him and for giving him a new start.

That interview ended up being an hour and a half long and perhaps it was the best ninety-minute investment in an employee that I had ever made. After several years, that man ended up becoming a model employee and eventually moved up to take a position as a supervisor. He kept his word.

There is a certain kind of person who sometimes just needs an extra chance in life. I was glad for whatever small part I played in getting that man back in the mainstream, and, over the years, it was just one of the many situations I ended up encountering.

People are basically good. My friend Joe Cucci used to tell me: "There are good people who sometimes do bad things, and then there are just bad people." But I like to think that, in general, there's more good out there than not.

It was my job to find those good people, and not just for the entry-level positions. The airlines were getting ready to expand their routes, and that meant hiring pilots. The company had not hired pilots in more than a decade, and I knew this would be a demanding, challenging task. I also knew what I wanted to do.

I committed myself to use this opportunity to increase the diversity of the company's pilots: women and minorities. The key was that there would be no lessening of standards or expectations. Diversity did not mean poor quality. Only the best pilots would be considered, and I always felt an obligation to have a good mix that would represent the profession.

We had an upcoming class of ten pilots to fill. I always made sure that not only was I bringing in candidates of different racial and ethnic backgrounds, but that I was including the different branches of the military service and civilian pilots, who had to work twice as hard to get time in the cockpit.

There were a lot of applicants, and there was a lot of work needed to weed out the best ones, not to mention the challenge of finding diverse candidates who met our standards. And, of course, I definitely felt that I had to adhere to my policy of posting the positions in-house as well.

It was a move that stretched the process out several weeks.

Little did any of us know who was out there.

Alberta Lee Parkison had worked for Texas International for ten years; she was a flight attendant with an FAA ticket in her back pocket. She also came from a family of aviators; her father was a retired Air Force B-52 pilot.

She had all the necessary training and certifications. In fact, she had trained eight Texas International pilots who had been her students when she was a seaplane instructor—and they were then flying as crewmembers on the DC-9, while she was serving the coffee on the same plane as the flight attendant.

When Alberta responded to the internal job posting, it was

too good to be true for me. You post, and you say to yourself: "I may never find anybody who can do *this.*" We were talking about certifications, licensing, and the proper number of flight hours—not commonplace things. But then to have someone step up and essentially say: "I can do this. I just never told anyone" was like a gift, a reward for believing in people and trying to live and work by ideals of equality and fairness.

And it was the beginning of a major breakthrough for the company as well. A culture that is exclusive and keeps certain elements of people out is tough to change but if someone doesn't step up and try to break the cycle, it never goes away. It just perpetuates and gets prolonged and entrenched as the years go by.

I was just so glad that Texas International was moving in this direction while on my watch. We had good talent and people—but if the chances I was taking had ended badly, I would have had my head handed to me.

The system as designed worked and helped discover Ms. Parkison among our own staff, but in general the process of hiring pilots to fly the DC-9 was long and tedious. Flying a DC-9 was a significant responsibility. The journey took me throughout the United States, and I was fully committed to going wherever I felt there were good candidates. I knew I wanted a cross section of candidates and spent a lot of time away from my family to ensure that we were finding the proper people.

This is when my previous life working in the space program and all those hundreds of hours I spent inside the cockpit of the lunar module came in handy. I've always been around flight crews and pilots, so I had a good sense of what I was dealing with. Pilots generally have big egos—and those were the ones who never really made it to second base with me.

Before I would go into an area or region of the country looking for job candidates, I would run an ad in the local paper first. Then I would set up an interview in a room set aside for that at a local hotel, one that was usually close to the airport so that I could fly into a destination without needing to rent a car or anything.

It was a process that took several months, and I was going all the time. No one told me where to go; I had the freedom to decide how to do this. I would often look at a cross section of the country—say Colorado, Arizona, and California—or I would look at where the flight schools were.

Often I would go to two or three states at a time doing courtesy interviews, and then I would decide who might be a candidate for a more in-depth interview. After a while, the names and faces would sometimes run together. I was talking to fifteen to twenty pilots a day, so you can understand how that could happen.

I would usually indicate on an application whether or not a candidate would require a follow-up. On one occasion, I must not have marked one of the applicants and had him flown in on a Texas International company plane, paying for all of his expenses to come to our corporate office. This pilot sat in my office—one of those egotistical guys that I didn't think would fit in with our corporate culture at all—and, now, here he was. I found myself having to do a courtesy interview, which took about ninety minutes of time that I just didn't have.

I thought honestly that I'd give him a second chance. Maybe my first instinct was wrong, after all. But sometimes your first instinct is the right one. Second guessing myself just wasn't worth it and I vowed never to make that mistake again. Besides, it was costly for us and the applicant.

From then on I would bring a Polaroid camera and take pictures of only those who were to get a second interview; then I would attach that to their application. It might not have been the most efficient way to do things, but it worked.

We were getting a good selection of candidates but I still wanted to meet our diversity goals. I was spending a lot of time with the vice president of flight operations, Don Breeding, and more with Robert Lemon, the chief pilot and director of flying, since the crewmembers I was hiring would be working directly with them and all the expenses of my job search was coming out of their budget. They weren't shy about letting me know, either!

Don was a hands-off manager who made sure you understood that what was important was getting the best candidates. Bob Lemon was a cool guy who would always look for a way to make you perform at your best. But he never lowered his standards or compromised on the quality of the candidates, and was probably the best pilot I would ever know. I would always trust his judgment, and would come to respect him as one of the fairest-minded individuals within Texas International. Bob was a very special person to me and I will always think of him in those terms.

I remember I was on a trip to interview candidates and it was late Friday afternoon. I had been gone for several days and by then really needed to get home to my wife and kids. As a member of management, I figured I would have a seat on a long flight home waiting for me.

The flight I wanted was full. It was Texas International policy to allow a member of management like me to bump a paying passenger and take the seat. But I didn't feel like that was right, to bump a paying passenger. I didn't have many options because it was the last and only flight out. So I did the next best thing: I called Captain Lemon and he gave me authorization to fly on the jump seat in the cockpit, which actually solved my problem.

As I settled into the cockpit, the captain and the copilot were talking and at some point during the flight, the copilot delicately broached the subject of TI's efforts to hire more diverse candidates. I encouraged them to express their true feelings and, wow, did I get an earful!

It seemed women working as airplane pilots bothered people more than the idea of working alongside others. I heard a lot of comments from these two pilots, including things like: "They are not as strong as us," and how women had "certain times of the month" that were worse than others.

For most of the two-and-a-half-hour flight, I stayed silent. But then I finally countered with some questions of my own.

"Do you have a daughter?" I asked. "What if it were your daughter who had all the FAA requirements, had secured the proper licenses and had the flight hours? Would you not want her to have the same opportunities you had?"

There was total silence. I think they finally got it at that point, and I was just drained.

It wasn't an easy time, by any means. But it was so crucial to do whatever it took to break through that kind of thinking—otherwise, how would we ever move forward?

In the end, we made history by hiring six women to be pilots at Texas International, including Parkison, the first flight attendant, and Jill E. Brown, an African-American woman who had been piloting planes since she was seventeen-years-old, as outlined in this 1978 Texas International press release from the time:

> Texas International Airline's six new female
> pilots received their First Officer wings along
> with 32 male pilots in special graduation cer-
> emonies... The addition of the six women,
> including the first black and the first flight
> attendant to qualify as pilots for a major U.S.
> commercial airline, increased the ranks of
> female flight crew members on U.S. carriers by
> almost 20 percent.

These women hadn't had it easy. None of the candidates to become pilots or first officers at Texas International did. They had to go through preliminary interviews, and then their applications would be looked over by a flight review board that would have to be set up in the corporate office. Only the candidates we felt we wanted to hire got to that point in the process, and the review board was made up of some heavy hitters, including Don Breeding, Robert Lemon, the vice president of personnel and myself.

It was especially stressful for me, since I would present these candidates that I had selected to the board. The candidates answered questions and interacted with the review board before they left the room, and then we would vote—thumbs up or thumbs down. For whatever reason, board members would have a hang-up with the way a candidate answered a question or something, and they would remember that when voting. While this was going on, we had the applicants wait in another office until their fate was decided. Sometimes it would take twenty to thirty minutes to discuss things, and I would come out and let the applicant know that we would be contacting them shortly. Then I would send them on their way.

I always tried to do this as quickly as possible out of respect for their feelings. It must have been sheer agony for these pilots, waiting in another room while their fate was being decided for them.

We'd reach out by mail shortly after the review board meetings. The ones we wanted to hire would be set up for psychological and military first-class background checks, as well as physicals. If they passed those, they would be offered the job of being a first officer on the DC-9.

Unlike those bitter days at Grumman when I was tapped to bring in the coworkers who were about to be laid off, these positive moments at Texas International were some of my most rewarding. From the process of first meeting these candidates to going through rounds of interviews and the physical, to being the one who got to call them on the telephone and make that job offer to each and every one—it was a great feeling. I was thrilled to make all those calls personally.

I never made them agonize more than I had to. I would get right to the point: "I'd like to offer you a position with Texas International as first officer, starting…"

Some people got quite emotional and were extremely appreciative. It was such a happy time, especially since I had spent so many hours seeking out and promoting these candidates. I'm not sure they really ever knew what an advocate I was for them. I always congratulated them for their efforts and hard work, and then cautioned them that they weren't quite off the hook yet and would have to successfully complete designated training before everything was final.

The doctor who handled the physicals for the review board was Dr. Gibson. He was the flight surgeon in Dallas who was in charge of screening the health of these potential employees.

After all, the health of a pilot who had hundreds of lives in his or her hands every day was key.

Dr. Gibson called me one day; he was always joking with me and trying to get me to take the same all-day flight physical that the pilots did. So this time, I took him up on it. I spent a day getting poked and prodded and living through what each of our job pilot candidates would have to go through, including the psychological testing.

It was all in good fun and good spirits, but toward the end of the day Dr. Gibson seemed to notice a lump in my neck. Things got serious. He immediately sent me for more tests on this nodule that looked a little strange, and it was determined that I had a growth on my thyroid. Dr. Gibson said I should have it looked at right away. He also suggested I go on blood pressure medication.

If not for his lighthearted joking, who knows what would have happened to me? I wasn't having symptoms so I had no idea anything was wrong. Eventually the growth on my thyroid was operated on and found to be benign, but I'm indebted to Dr. Gibson for finding these problems early on. He was great at what he did, and I had a lot of respect for him.

I am lucky to have known so many good and wonderful people in my life. I once heard it said that behind every successful man is a surprised mother-in-law. Well, that was never the case for me because I have a mother-in-law who is one-of-a-kind.

Mabel Suddin is only about four-feet-seven-inches tall, and all of ninety pounds. She is one of the sweetest and most caring people I know, and has always been encouraging. One of her typical sayings that quickly became a favorite of mine is: "David, you're so smart, and God bless you."

Defying the stereotypes of bad mothers-in-law, Mabel was

never the interfering type. A strong, dedicated woman from the Bronx in New York City, Mabel raised eleven children, including my beautiful wife, mostly by herself.

She is a rarity—trust me, they don't make that model anymore. I'll never forget her first visit to our house in Houston. When Kumar was young we didn't have babysitters, so my mother-in-law agreed to watch him for us when we were going on a trip to Hawaii.

Mabel didn't drive, so we stacked the house with food and supplies and got everything as ready as we could for her. We flew her down to Texas—I'm not sure she'd been on too many airplanes—and she was really nervous, especially being in an unfamiliar house and not knowing the area.

One night the wind was really whipping about, and Kumar had been sleeping. I don't know if Mabel got scared and woke Kumar up, or if he got out of bed on his own. But as the family story goes, the noise of the wind rattling against the windows was unnerving. So she asked Kumar what was making the noise.

"Oh, Grandma," he said, "it's okay. It's probably just someone trying to break in."

I couldn't have asked for a better mother-in-law. She is the kind of person who turned out to be a real treasure and my love and respect for her is endless.

Little did I know it, but around this time I would stumble across another person who would turn out to be a source of pride. He was a job candidate at Texas International whom everyone seemed to like, and whom I really hoped we would be able to hire. Buddy Howard was applying to be a pilot with Texas International, and if everything worked out, he would be the company's first male African-American pilot.

Buddy Howard was the ultimate professional pilot. He had

flown T-39s in the U.S. Air Force and had several thousand hours of flight experience, along with all of his FAA ratings and certifications. He passed all the requirements, the physical and background checks. That's when I started to hear conversation around the office—general talk about how some ethnic groups have a tendency toward elevated blood pressure or end up needing glasses.

The talk made me angry. What did those generalities, which were probably based more on stereotypes and discrimination than anything else, have to do with anything?

I knew Buddy could fly a plane, but I wondered if he'd be able to function, be able to do his best work, in the face of this good ol' boy network that was still alive and kicking at TI. In those days, they had just upgraded to the DC-9, and our current pilots would not have as much technical training or professional experience as the candidates that were being considered. Usually, the new people would be assigned menial chores in the cockpit, such as giving the announcements. That could be a tough thing to deal with.

Originally we had decided to fill a class of ten pilots and would hold over some remaining candidates to call in for the next class. Buddy was one of those, and I was disappointed.

The class was cut to nine people—that's how many I was allowed to hire. It was frustrating to be on the edge of this great opportunity, this chance, again, to make history in our company and to break unjust barriers, and we weren't taking it.

I had been given the green light for nine people, but I was thinking about Buddy. He had been so close. Then, I made a potentially career-altering decision.

I called Buddy Howard and offered him a position as a first officer, making him the tenth person in that class of pilots.

When the time came later for them to start training, all ten were processed and immediately went to class. The next day I was sitting in my office when I heard a knock on my door. Don Breeding, the vice president of flight operations, slowly opened the door and poked his head inside my office.

"How many people were in class?" he asked me.

"Ten," I said.

Don stood there for a moment, quiet. In that moment I could feel my heart beating fast and prepared myself for a hard time.

"I thought we agreed on only nine slots. Who was the other pilot?" he said.

"I offered Buddy Howard that tenth slot," I said, slowly exhaling.

He looked me straight in the eye and said, "We'll see what happens."

Then he shut the door. I knew this was it, a pivotal moment in my career. I ran into Buddy a few days later and pulled him aside. I emphasized that he was paving the road for others to follow and what he needed to do—I'm sure it was overkill on my part. Buddy was a smart guy who knew what was at stake.

Months later, I was on another recruiting trip. I was in a conference room at the Dallas Airport when I ran into Don. We were talking when we looked out onto the tarmac. And there, out of the corner of our eyes, was this proud African-American pilot, with his immaculately pressed uniform, carrying his flight bag and walking toward the airplane he would be copilot of that day.

Pride surged through me and I couldn't help but smile. But Don finally put my insecurities to rest in a moment I will always remember.

"What a qualified pilot. You made the right decision," he said.

After that, I was never questioned on budgets or candidates ever again. I was to go on many more recruiting trips for pilots over the years.

It seemed everyone I ran into had someone they knew who wanted to work for the airline. Many times I wouldn't tell people where I worked or what it was I did, even employees and management. It was hard to get people to understand that we actually had a process that we stuck to. I was often tested.

One day, my secretary handed me a pilot's application. There was a handwritten note attached.

"This came from a Congressional member. Dave, please handle."

This note was from CEO Frank Lorenzo himself.

Just what did he mean by "handle"? Did that mean to find a slot for this person? What did that mean for the process we had worked so hard to establish?

I was given no more guidance beyond that note, so I made an executive decision and asked my secretary to set up an interview in my office with the candidate. This meant that unlike my other recruiting ventures, I'd be footing the bill for this candidate's interview out of my own division's budget.

A week later the applicant was sitting in my office. He had a big grin from ear to ear, and I knew he was thinking that this meeting was just a procedure he had to get through before moving on to the next level. I spent an hour reviewing his background, education, flight hours and the equipment he had flown. At the end, I thanked him for his time. Then I sent him on his way.

A follow-up letter was sent the next week, letting him know

we had no pilot classes scheduled, but that we would keep him in mind.

I hoped that meant the matter was finished.

Several weeks later, I was walking along the halls of the top floor of our office building—we used to call that area Mahogany Row because all the top brass had their offices there. I saw Frank Lorenzo walking toward me, and I knew he'd ask me about that applicant.

As Frank got closer, I could see myself standing on the unemployment line.

"Hi, David," he said, and walked on past me.

What a relief! Until he stopped and looked back at me.

"Whatever happened to that pilot applicant I forwarded to you?"

I took a deep breath.

"Frank, he had the flight hours on good equipment, but I wasn't sure his demeanor or personality would mesh with our corporate culture. Frankly, I think we can do better, but if you want me to…"

He stopped me before I could finish.

"Okay," he said, and kept going down the hall.

I had been nervous, but that encounter with Frank was symbolic of most of my time at Texas International. I was never pressured or asked to consider a candidate if I felt they would not be a proper fit for the company. I had proven myself, and no one challenged me—even though I always knew the veto power of the other managers was alive and well.

I was so proud of the diversity that took place during those years. While I would be employed at Texas International, the infusion of female pilots, and the hiring of Buddy Howard, would always be among my most cherished moments.

Giving people a chance to realize opportunities they may not have otherwise taken advantage of was a thrill for me. I feel strongly that everyone has a chance to find their own kind of success. Not everyone has to fly the plane in order to be on top of the world.

In those days I was quite active in the Houston community, my adopted hometown. I was always out encouraging and seeking out new and creative talent. After all, there were hundreds of jobs within an airline like ours, and room for everyone to achieve their dreams. I had participated in a task force committee that had been convened by Vice President Walter Mondale's staff, and was not afraid to share my input on youth initiatives and employment. I often spoke my mind—maybe that's how the New York directness came in.

Several months after my work on the committee, I was working in my office at Texas International. It was a typical day. My secretary, Susan, was date stamping hundreds of applications our office would receive in any given period of time.

When she knocked on my office door and walked in, she looked strange. Susan didn't look like herself. Then she said: "There's a mailogram from the social secretary of the White House."

I was taken by surprise. Susan kept talking, slowly explaining that President Jimmy Carter wanted me to attend a function on January 10, 1979, in the East Room of the White House, with a reception to follow in the state dining room.

Now, Susan had my attention. I stopped what I was doing and just stared at her for a moment, not really believing my ears. Me? At the White House?

I stopped the work I had been doing and began to get the information together that I would need to submit for the required background check that would be done by the Secret Service and the Federal Bureau of Investigation.

I think I was in a bit of shock. When I mentioned the invitation to my boss, I think he and the rest of the management at Texas International were more excited than I was. At first, my emotions were slow, like molasses, but as time went on I couldn't hold back my excitement.

I tried to be cool as I nonchalantly explained to Amina how my day went. By that point, I had a hard time holding back my growing enthusiasm.

Hotel and flight plans were made and now—the hard part. Waiting for the day to arrive. I was like a kid waiting for Christmas morning. The anticipation was tangible, and as the day of the reception got closer, my anxiety increased.

As my nervousness increased, the days seemed to get longer, until finally it was the night before. I made my way to Washington, D.C., checking in to a hotel close to the White House. I think it may have been the Mayflower Hotel.

The evening passed in a haze of anxiety, anticipation, and wonderment. I made sure the suit I had picked out for the big day was neatly laid out. I grabbed a quick dinner and decided to stay in for the night, so I could be well rested.

Sometimes I oversleep. That was not going to happen on the day I was going to be visiting the White House. I had two alarm clocks to help me wake up, and then made sure that, as a backup, I had the hotel give me a wake-up call.

I must have read the letter and invitation more times than I wanted to admit. As the hours ticked by I found myself practicing what I would say to the cab driver that would take me there the next day.

"Good morning, the White House, please," or, "Good

morning, take me to the White House, please" both had a nice ring to them.

Here I was, a corporate executive from Houston, about to attend a reception at the White House. But I was so much more. I was someone who had helped, in my own small way, to send our country's astronauts to the moon.

I was a father, and a husband, and I was planning to start my own business soon.

And I was still, deep down, a kid who spent his summers working in a junkyard, who had to walk miles and work so hard just to finish high school.

Here I was, with an invitation from the president in my hands, and a White House reception in my near future. Wow.

I don't remember who told me this, but it's been said that a person shouldn't be judged by how far they've come, but by how far they've come as compared to where they started. You can't go too much farther from a Long Island junkyard to a presidential reception. I was feeling pretty good.

For once, I woke up early and was on time. I hailed a cab, and in my nervous state muttered my much-practiced words to the cab driver. But he got it, and took me directly to the White House.

When I arrived I gave the security guards at the entrance my information, and a very large individual in a dark business suit took it and looked at what I had submitted. Then he looked at me; it seemed like forever as he looked me over, top to bottom. Then he shook my hand. He motioned with his left hand as if to point out the curved driveway that led up to the front door of the White House, and said:

"The president is expecting you, Mr. Cisco."

What a feeling! I was practically floating as I walked up that driveway. On either side of the front door, there were two

Marines in uniform. Suddenly, the thought hit me: "I wonder if there is another Dave Cisco? Do they have the right one?"

But that disturbing thought passed quickly. I didn't know what to expect on reaching the front door. I was greeted and when I walked into the White House, there was a small military quartet playing chamber music.

I walked through the foyer, viewing the paintings on the wall. I remember the large image of John F. Kennedy—the president who launched the space program. I even stopped and had my picture taken in front of it. I had a chance to view several rooms before the ceremony. It shows how different things were back then than today.

I got to sit in the presidential chair that once belonged to Thomas Jefferson. I looked back behind me and asked one of the Secret Service agents to take my photo. I was like a kid in a candy store, marveling at everything before me.

I sat down in the Red Room of the White House and again asked someone to take my picture. It was so quiet and relaxing. I took a few minutes to absorb the peaceful atmosphere. It was just a few minutes but it seemed like forever, sitting there by myself. No one was around, but it felt like there must have been cameras watching what I was doing.

It was a time in history that we were fortunate to have, but one that we will never see again. There's probably no way that could happen again, given the heightened security brought on by the September 11, 2001, terror attacks.

At the ceremony itself the president announced his initiative on youth employment. I was seated in a room with other invited guests and, after the ceremony, was taken to the state dining room, where the reception was held. I was so wowed by the whole experience, so excited, that I had an overwhelming urge to take a piece of the day home with me.

This compulsion to snag something from the White House was strong but there was a memory that I couldn't get past: my mother and those cat's eye marbles. So I thought better of it, and instead I settled for some napkins with the White House logo on them. Surely that would be all right. It would be a token of the day, however small, and one that I would cherish for the rest of my life.

This was a highlight of my life. I was bursting with pride as I returned home to Houston, to my family and my work back at Texas International.

I was doing well at the company and enjoyed my time there. I felt I had accomplished so much and helped the company achieve important goals. But I never chose to move up the corporate ladder. Many members of management never understood why I hesitated, and why it appeared that I had passed up so many opportunities to move farther along within the company.

But I had my reasons. For someone else, this job at TI might have been the pinnacle of a career, a crowning achievement. But I had my sights set on something different. I knew I had promised Amina that this was a temporary job; we had committed ourselves to opening our own business, and Amina had already started working toward that goal. I would eventually join her.

We named our agency Bay Area Travel, Inc., and opened it in January 1979. Amina had been working there from the beginning, while I had been doing marketing during my time off.

Before we opened, I had applied to Nassau Bay Bank, a local institution, for a loan of $10,000. I wanted to make sure we had adequate capital, and we wanted an office with good visibility. We had submitted a packet with all our financial papers, along with our business plan, weeks earlier.

We were confident in our plans and goals, so later when I got the letter rejecting us for a loan—on the basis that we lacked experience in our chosen field and that there was no room for another travel agency in the Bay area—it was devastating.

The disappointment was hard to swallow, especially when we hadn't really gotten off the ground yet. But Amina and I talked about it, and were determined to continue despite the setback. We decided to take a chance and go ahead without a loan. We capitalized the business ourselves.

There was a small retail storefront in Webster, Texas, at 1558 Highway 3, with 1,000 square feet for lease. It looked like a great opportunity, with just one catch: We had to sign a five-year lease.

This was a huge commitment, made even more daunting by the bank's refusal of our loan application.

When the lease forms were placed in front of me and the pen was in my hand, I was frozen. I was so nervous I don't think I made any move to sign that paper. The pen seemed to jump around in my hand and, before I knew it, my signature was there on the lease form.

Now, it was done. We'd taken a chance, and hadn't done so lightly. Some might have thought it silly to be risking so much when I had such a good job already. The business, though, was like a promise we had made to ourselves to be our own bosses and to forge our own path.

We had taken these steps forward together but, really, it was Amina who got the business off the ground. One day, the moment of truth came. Amina took me aside and gently told me it was time for me to consider coming into Bay Area Travel full-time, and that we should be getting aggressive in seeking new accounts.

I knew she was right. We talked about it for a few weeks but there was no way to avoid it, and while I was quite satisfied with my time at Texas International, leaving had always been a part of my master plan. I had already made that choice. However, deciding when to leave wasn't an easy thing to do and I could feel the stress building inside of me.

I had shared my views with Don Breeding and Robert Lemon, both of whom I trusted. One Monday, I finally tendered my resignation at TI, and once I did, I felt a lot better. Just making the decision public took a lot of the weight off my shoulders in a way.

After all, a great deal was accomplished during my tenure, and now it was time to move on. Everyone was sad to see me go, but they understood why I was leaving and finally understood why I had never taken a new position that would require me to relocate, or one that moved me up the company ranks.

Frank Lorenzo, the CEO, was leaving the country and wanted to talk to me before he left. As I walked up to Mahogany Row, I kept saying to myself, over and over: "I will not be talked out of leaving. I will not be talked out of leaving."

I walked into Frank's office, and we had a nice talk about where we had each come from, and where TI was headed. He assured me the company was going in the right direction, and he wanted to know why I was moving on.

I took a deep breath, but didn't really hesitate. I explained to Frank that my leaving really wasn't about Texas International. Rather, this was something I had to do, for myself, for Amina and for our family. I knew he understood.

We shook hands, and that would be the last time I ever saw Frank Lorenzo. I processed out and said my goodbyes.

Over the years I had forgotten just how many people had come through my office. Each one had a different story. Once again I was bursting with pride at the history the company had made, at the corporate culture we had created that valued the best people with the best skills and given opportunities and chances for success to those who hadn't previously had them.

As I was leaving Texas International on that last day, my emotions began to change like a roller coaster, as pride mixed with nostalgia and a little bit of sadness. It was getting dark outside as I walked the several blocks to my car in the company parking lot.

It was also misting out as I walked, coating everything with moisture. With each step, reality began to set in.

I had just quit my job.

I had two small children, a mortgage, a new business—there was no medical insurance or any guarantee of a paycheck at the end of the week.

Even though I knew what I was doing, had talked about it and planned for it and worked for this change for years, I think I went into a kind of shock. The mist and the dark hid the tears that seemed to suddenly spring from my eyes and stream down my face as my emotions intensified.

It was a short walk. Those few steps, though, were some of the most important ones I ever took. As I walked I became more intent on becoming successful on my own, on making Bay Area Travel work and grow into everything Amina and I wanted it to be. Amina had already worked so hard, and this was something I had hoped for since Tom Short first helped me to see that it was possible. Now, as I took those first steps into my new life, I swore I would do whatever I could to make the dream come true.

Just because I willed it so didn't mean it would happen,

and certainly not right away. The transition from airline company executive to fledgling entrepreneur was a little jarring. At Texas International, my phone used to ring off the hook. I had my own office and a secretary. I would always have a stack of messages that needed to be returned.

Now, in my first days working full-time at Bay Area Travel, no one was waiting to see me. There were no messages. One day, my phone never rang at all, and I even had someone call me just to see if it was working properly. It was. Talk about a deflation of ego.

It's so hard when you're a small business trying to establish yourself. It's key to market a new venture, get the company name out and try to drum up business, but the first questions are: "How long have you been in business?" and "Who are some of your accounts?"

While I focused on getting new clients, Amina dealt with business operations and making everything happen. Believe me—I had the easy job.

Things started to come together slowly and Bay Area Travel began to establish itself. I believe Ford Aerospace was one of our first major accounts, and then after successfully handling their corporate travel, we responded to several requests for proposals from IBM. We bid competitively and were awarded the contract.

That was with a year and a half of business under our belts, and we must have been doing something right because everyone was pleased with our services. At that time, we decided it was time to really branch out and expand, so we opened a second office.

The downtown office of Bay Area Travel was at 1000 Louisiana Street in Houston. By this time we seemed to be on a roll, proving the downtown office was a good idea. We secured accounts like Lyondell Petrol Chemical and Arco, and were handling the travel arrangements for the City of Houston, the Mayor's Office, and a host of others.

We were lucky, to be sure, but our success in the travel agency business had more to do with the hardworking, dedicated people who chose to work with us. I always knew that in order to be successful in business you had to have like-minded employees—and after all, I was pretty good at finding qualified candidates.

Pat Debord was our bookkeeper and the person in charge of our accounting; Tonya Lasagna was the manager of our office in Clear Lake. Mitzi Almeida would become the manager of our downtown office. These were loyal staff members, and they grew with us and made major contributions to our success. Bay Area Travel wouldn't have been the same without them; they were family.

Before we knew it, years had gone by and Bay Area Travel had done good business catering to the travel needs of Houston's corporate community. We had proven not only that we could do this but also that there was indeed room for another travel agency in Houston. But as the years progressed, we knew we needed to make a change to increase our bargaining advantage.

It wasn't too much of a surprise when American Express approached us. After all, Bay Area Travel was doing great volume in two offices. But I felt we needed to bump things up to another level. So we went into discussions with American Express and eventually decided to become an AMEX representative office.

The great thing about this move was that we would still be an independent business, and we would still call our own shots, but we would be representing the AMEX brand and selling their products exclusively, and that gave us additional credibility.

The hard work never stopped. Every account we marketed to involved a long and tedious process of writing proposals and putting in bids that showed how we could reduce the corporation's costs for air travel, hotels and cars, all while delivering 100 percent satisfaction.

And we learned pretty quickly that we had to do all of this while also doing everything the right way. If we made a mistake, we'd pay for it big-time. From time to time, some of our accounts would hire third-party organizations that would randomly audit our work. If they determined the lowest rate was not applied, we would have to pay the difference, as well as a penalty on top of that.

It only took a few of those penalties to realize it just wasn't worth it. When you're in a rush and worried about just getting more business, it seems like there's never time to do it "right." But when things go wrong, there's always time to do it over. So we learned quickly that everything had to be exact. Every staff member knew it was better to "do it right" because, if not, you'd have to answer for it later.

Bay Area Travel continued to grow year after year. We kept getting new accounts such as Lockheed Martin, Computer Sciences Corporation, Amaco Chemical, AT&T throughout Houston and Dallas, Motorola, Boeing, Anheiser Busch, and many others.

Eventually we decided to make a change with our downtown office and renamed it Airline Travel Service. We operated both of our offices independently for tax purposes, and both did well.

As Bay Area Travel grew, Kumar and Lahna did as well.

I had always wanted our children to go to college and to find their own paths in life—and they did.

Kumar graduated from the University of Washington in Seattle, and Lahna went to and graduated from the University of Texas. Both secured corporate jobs.

Lahna, in fact, was on her way to breaking glass ceilings with a corporation in Dallas. I remember going to her office, a beautiful high-rise building in downtown Dallas. She had a gorgeous office with her name on the door. Now, how cool is that for a father? I even secretly took a picture of it. I know that sounds a little anal, but, well, no one saw me do it—and *that* was a good thing because I realized she would be very embarrassed if they did!

As time went on, Kumar, who had been working in Washington, was considering a change. He was thinking of coming back to Texas. Amina and I were thrilled and thought it would be great if both of our kids were back in the Lone Star State.

One night after dinner, Amina approached me with an idea.

"Why don't we offer Kumar a position in Bay Area Travel?" she said.

No, I thought. My first response was that I honestly didn't think that would be a good idea, given how tough it can be to work with family members.

Granted, Amina and I had been working together in the agency for many years, but I usually worked out of the downtown office and would only come to the Bay area office maybe two days a week, so we weren't exactly sitting next to each other all day.

Still, Amina was persuasive, and she and I and Kumar had many conversations on whether or not Kumar should join us in the business. We agreed it could work, and we offered him a job overseeing the accounting department. He also offered his technical abilities so that we could automate some of the redundancies in our day-to-day work and increase productivity.

I was often regarded as a bit of a tough parent—though always with the goal of helping my kids to become self-sufficient. This situation was no different, and Kumar started with a mere $1,500-a-month salary. I wanted him to earn his stripes.

He worked in the downtown office with me, and we would travel together to the Bay Area office. We agreed to evaluate his work after six months, but he immediately fit in and made and implemented many new and creative programs that contributed to the agency's success. It was a pleasant way for things to turn out, and it wasn't until later on that I realized why.

Kumar became a key part of Bay Area Travel, and after a few years he pointed out to me that he was just seven years old when we started the agency. He had grown up listening to Amina and me talking about what we needed to do and how we could make the business better. So while I had at first thought maybe bringing our son into the business wasn't such a good idea, I couldn't have been more wrong. He was a natural fit and bringing him in was one of the best decisions we made.

With Kumar's help, we were able to take a leap forward in how we did our work. New technology meant we could take customers from all over the country. That became key because as time went on we shifted our focus from leisure travel to mostly corporate accounts, which were much less labor-intensive and quite profitable.

Since we were getting a ten-percent commission on all tickets booked and many of these corporate accounts flew first-class all over the world at an average of $5,000 a ticket, we did quite well.

At one point we were rated by the *Houston Business Journal* as one of the top Houston travel agencies specializing in business travel. We had twenty-eight employees between both of our offices and were doing approximately $28 million in business a year. Each major airline was seeking our favor in hopes we would direct more business their way—and many couldn't figure out how we were doing so well.

How sad. The secret of our success wasn't any big mystery, as far as we were concerned. It's based on the idea of hiring the best people, only qualified talent, and treating the clients exactly how you would want to be treated. I know it sounds like a cliché, but the thing is it's true.

One of the fringe benefits of having such a successful travel agency was that it allowed Amina and me to travel much of the world, and to do so first-class. At one time, because of the amount of business we directed to Continental, we were issued a Continental air card that allowed us to fly at any time, first-class, and without a reservation.

By the mid-1980s, Bay Area Travel had made it. We were a success, and all the hard work and risk Amina and I had taken in the beginning proved worth it. But we realized it was time for a change. We were too big, doing too much business, for this location.

So we found something different just off NASA Road 1: a small, 2,400-square-foot building that we bought in 1984. We now had our own corporate headquarters. We kept the downtown office open and maintained Airline Travel Service there, but moved Bay Area Travel to the new building.

To think about the chance we took, starting out with no financing and nothing but our own hard work and ambition, and then to realize we ended up doing so well in just a few years, was breathtaking.

One of the first things I did was have a thirty-foot flagpole installed in front of the building. We flew the Stars and Stripes from it, and hearing the flag snap as the wind blew made me feel somehow bigger, as if my soul were expanding.

I would always look forward to the sound that flag made and how beautiful it looked. It seemed to be a real-life representation of how I felt. I was truly living the American dream.

I'd sit in my office and look out of the window at that flag and know that I had made it.

Nowhere else can people rise so far, or achieve so much, if they choose. It's one thing to tell stories about how people go from rags to riches or from obscurity to fame. But the how and the who almost don't matter. What matters is that here, if you apply yourself, nothing is impossible. This is the best country in the world—for everyone. You can become independent, you can gain wealth and you can build great projects or accomplish historic ends. A person can succeed in America at any venture they choose. The opportunities are there for the taking.

Sometimes, it doesn't hurt to help nurture those opportunities. If it weren't for the different people and institutions in my life helping me along the way—my mother, Mr. Nobman at the hardware store, the chances I got at Grumman—things might have turned out very differently.

I've always believed in paying it forward and doing whatever I could to help someone less fortunate, even if it was just by doing so in my own small way. For me, it was a question not just of doing what was right, but of repaying some of the kindness that I've known over the years.

It was that instinct that made it easy for me to say yes when the CEO of one of our corporate clients at Bay Area Travel asked me to give my time and attention to an organization called the Houston Education Committee. The name of the group was later changed to the Houston Business Committee of Educational Excellence.

This board was made up of the movers and shakers of Houston. We're talking important business types who ran successful corporations and who were well-known in the community. They were all heavy hitters, except for one—me.

In a way, it was like being back at those weekly management meetings at Rockwell. I definitely felt a bit out of place among all those people. But it was one of the most gratifying experiences I had, and made me feel like I was a vital part of the community doing something that had a great impact.

The goal of the committee was to motivate the area's teachers and to help them. We would raise money, donate our time and support teachers who needed either funding or assistance with special projects for their students. We ran an awards ceremony each year recognizing the Teacher of the Year.

I was often singled out to go and talk to students about how I became a business owner, and how important it was to stay in school and get a good education. I had no problem being put in the hot seat, so to speak. For me it was a special, important mission. I enjoy working with young people and I hope people find, in my story, their own motivation to succeed.

At each year's ceremony, our organization would announce the Teacher of the Year, and the nominees and their students would attend a banquet that sometimes had as many as several hundred people in attendance. The committee would give the winner $5,000 to use however the teacher wished, and our company, Bay Area Travel, would offer two free airline tickets—usually to Europe, or anywhere the teacher wanted to go.

What I've learned over the years is that no one can do it all, and it's okay to try, in your own small way, to help the next generation, even if that means helping one kid at a time, or one teacher at a time. Really, in a way, it's a question of quality rather than quantity.

It's unfortunate that some people who work with children every day get burned out, or never realize how much power they have to help people. Teachers, school counselors, coaches all have tremendous influence. All some kids need is a kind word, a bit of encouragement, or just someone to listen to them.

My parents were good people, and they wanted the best for us. But they were worried about keeping a roof over our heads and food on the table. They didn't have the resources to send us to college, or even, really, an awareness of what that meant.

Back then, I would have loved to have someone looking out for me. I know what it's like to be that struggling young person, to feel lost and to have a seemingly dark future stretch out in front of you. A willing ear, a person with a positive attitude who helps you to have faith in yourself—that's priceless to a child in need.

If someone singled out students who showed a willingness to learn but who were disadvantaged, or who had trouble at home or lived in a rough neighborhood, and just offered some help—even tutoring, if that was what was needed—then just how many people could be kept from falling through the cracks?

I managed to catch up, but I did it on my own, and it's always harder to do so when you're already starting from behind everyone else.

It's all about how you take the time. Think about those cat's eye marbles. My mother could have spanked me and been done with it. She could have yelled and screamed. Instead, she was quite deliberate and took the time to teach me something. If she hadn't, maybe the whole thing would have just been marbles. Instead, it became so much more than that.

Everyone can become a success, even though some people start at a greater disadvantage than others, or end up taking a few detours along the way. But almost anyone can become a top ten-percent success story—it doesn't matter if success comes quickly or gradually, but it can happen.

What's sad is that the number of kids who need help is so great. But you can take a child and put him or her on the right track. The earlier intervention begins, the better. No matter how much I want to help people, no matter how much good anyone seeks to do, I've come to accept the simple truth that it's okay, maybe it's even better, to work with one or two kids at a time. If you can give of yourself just that much and contribute, monitor, nudge and be a part of someone's life, you can move mountains.

Reaching out to those around, doing the right thing, contributing to the community—these were qualities and habits my parents modeled for us whenever they could. In 1988, one of the saddest moments of my life was when my father passed away.

He was seventy years old. Over the years he had been plagued by so many obstacles, including illness and money troubles that endangered his family. But my father left this earth knowing that his six kids had gone on to make it. He knew everyone was not only doing fine, but thriving.

My brother Floyd, the oldest, the favorite, had served in the U.S. Air Force and, like me, worked at Grumman for thirty years and then for another fifteen years at Northrop Grumman. Combined, he had a forty-five-year career before recently retiring. My sister, Elaine, worked and then went on to be a housewife with her own family to raise.

Ronnie, the mischievous one, was a soldier in the U.S. Army for twenty-six years and worked thirty years for IBM.

Wayne also served in the Army, and, like our father, worked on the Long Island Rail Road, where he was employed for thirty years.

And my youngest brother, Skip? He had proven his status as the smartest of the bunch. Skip had earned a music scholarship for college, as well as one for athletics, but also played semi-pro basketball. At just five-foot-nine, he realized at some point he probably needed a better career path, and went into financial planning and insurance. Like me, he started his own business, with his wife, Sandy, in Seattle. It would become very successful, and I'm so proud of him.

When my father died, I made a pledge to myself. I vowed that I would return every dollar that had been given to my family while we were on welfare. Two years later, I started down that road and with the help of my financial advisor, Les Butler, established an irrevocable charitable remainder trust through a probate law firm. This wasn't something I advertised; rather, it was a private promise I made to myself. I didn't even tell my family.

The fund is not something that can be undone, and it is currently valued at approximately a quarter of a million dollars, which far exceeds what our family was given, though I can change which organizations benefit from it. It's a vehicle to make sure the funds go to the right place, whether that's the Boy Scouts, St. Jude's Children's Research Hospital, or our returning military veterans. For me, it was a choice but also a responsibility, a need, to give back the kind of help that had enabled my family to survive.

After all, why shouldn't I? Maybe it's the kind of thing that ends up helping another family. I wanted my mother and father to know, in some way, that, yes—we made it. We all did well. And every penny our family received for assistance when it was needed has been repaid many times over. It's all part of life's great cycle that, if you can, you repay the resources, help and love that you yourself have received.

I work to give back anonymously by donating, or by helping kids one by one. That's what I try to do. It's about doing the right thing. I don't know who came up with that, but it's true. My commitment is to keep my eyes and ears open and look out for that individual who is like me. There are many.

It's not easy work. Kids today are in so many ways different creatures, born of a different age in which they're more likely to dream of being rap stars than holding down an after-school job. The trick is to get a young person in need to understand that they can make it, that there are so many doors that could open for them, as long as they get the chip off their shoulders and apply themselves.

Over the years I have done much of my charitable work anonymously. I do this for a few reasons and it's worked out better for myself and my family this way. But, really, it's because taking credit for this kind of work isn't the point. Someone once told me to approach it as if your mission was to do something kind for someone today but not let anyone else know. If anyone found out, the kind deed wouldn't count. I have followed those words for many years.

The Teacher of the Year awards weren't quite so secretive. But the wonderful thing about the Houston Business Committee of Educational Excellence was the way we were able to help a greater number of people. The committee has awarded more than 1,100 "mini-grants" to teachers for innovative classroom projects, forty grants of up to $3,000 for principals to use in addressing school-wide concerns, and, in all, donated more than $350,000. We also paid tuition and housing for sixty local school administrators who went to summer school clinics at

Harvard University or Texas A&M, and gave annual outstanding progress awards to students.

While I did charitable work for many years, and still do, my work with the committee was extremely gratifying. It was being a business owner that opened that door for me, and allowed me to be a part of it.

It's an example of how great, and unexpected, the rewards can be when you challenge yourself and take a chance. Tom Short was right, and by encouraging me to branch out on my own he gave me perhaps one of the greatest gifts I've ever gotten.

The travel agency wasn't the only business Amina and I tried our hand at. That corporate office we had bought for ourselves had a lot of empty space on one side, and good talent was hard to find. We decided, in later years, to launch a training school to help people learn the ropes of the business.

It was a great idea, again, born of simple principles. If you want to find good employees, the best way is to train them. If you want people to have the chance to work in your industry, give them a place to learn it.

The venture gave us good employees for many years, and was quite a profitable endeavor. Making the effort showed how few limits there really are for people who want to go far.

Kumar stayed with the agency for a decade. He helped us move ahead with technology and was a huge asset for us. By the time he left to move on to another career, he was making close to a six-digit salary, with his bonus—reflective of his value to the company. I truly treasured Kumar's knowledge and commitment, and was fortunate to work together with my son for so many years.

I've learned over the years that what makes a man who he is, is not where that man starts out, or even, really, where he ends up. It's more important to realize how far a man comes from where they started.

The momentum I gained by leaping from a safe corporate job into the unknown waters of being an entrepreneur took me farther than I ever imagined.

By then Amina and I were living in one of the most beautiful neighborhoods in Clear Lake City, a suburb of Houston near the space center.

The streets were lined with large, stately oak trees and wonderful homes in Clear Lake Forest. My lawyer and friend, David Feldman, and his wife, Peggy Feldman, lived only a few blocks away. We had all moved to the neighborhood around the same time. Their sons, Chris and Seth, played together with Kumar and Lahna, and they all grew up together. Chris would go on to become a prominent Houston lawyer and Seth, a marketing guru.

David Feldman had run his own law firm, and later became a partner in a major downtown Houston firm. So, of course, they had the biggest house on the street. Our families are quite close; Amina is even Seth's godmother. So when he asked me to do him a favor, I didn't hesitate.

Then he told me it wouldn't take too much of my time.

Talk about underselling.

He wanted me to finish his term as a board member for the Clear Lake Forest Homeowners Association.

After all, I thought, *I do live in the community. I should be willing to give something back, right?*

So, a short while later, the other board members for the Homeowners Association voted me in, and there I was. I didn't just finish out his term. Instead, I found a niche with the association and stayed with it year after year. After a while, I was elected as the group's president. Running your own company makes you pretty results-oriented, and I wasn't shy about enforcing the deed restrictions that governed our subdivision.

Enforcing the rules is never a pretty job, and the first step was often to just have a conversation with the homeowner explaining the problem and why the major infractions had to be addressed. But I learned something: It goes a lot easier when you just explain the situation first, rather than sending a poison pen letter or making threats. Everyone, after all, wanted to improve his or her community, so we all had the same goal in the end.

Not everyone was so easygoing, of course. There are always a few people who see themselves as being above the rules, but I dealt with that by going to the justice of the peace, on my own time, and bringing the case to court. It was a deed restriction that had been violated for many years, but the judge ruled in favor of the Homeowners Association, and that was quite satisfying.

Then, one day, I got a phone call from a local city councilman. He was an official of Taylor Lake Village, the lakefront town that our subdivision was in. He asked me to go to lunch the following week.

We met in a local restaurant, and he introduced himself and another city councilman as well. By now, my curiosity was up, and with my New York directness intact after all those years of Texan living, I broke the ice pretty quickly. It didn't take me long to ask them what this was all about.

They were impressed, they said, with the work I had been doing as president of the Homeowners Association. Then he asked me if I would consider running for a position on the city council.

To say I was surprised was an understatement. I had never thought about running for elected office; I was just working in my own community, keeping the business going, doing my charitable work. That had been enough for me.

I'm not sure how well I hid my astonishment, but I thanked them for their offer, and told them I'd give it some thought after, of course, I had discussed it with Amina.

So we talked about it for a little while and tried to weigh the pros and cons. I was a pretty busy guy to begin with—would I have time to devote to something like this? But why not? I mean, how often does a chance like this come along in life?

Taylor Lake Village is a small community, where many of NASA's astronauts lived. I figured why not go ahead and do this, so I went through with the process of running for public office. Mind you, I didn't do very much to promote myself. There were two other people running for the council opening that year, and I wasn't exactly putting on a hard-fought campaign.

I don't know whether it was word of mouth or something else, but the election was held in May and, somehow, I won it.

Now my mother really had something to secretly boast about! She had once congratulated me for sending a man to the moon and had supported me through the years as I joined corporate America and then as Amina and I launched the agency. This was a whole new level. I think she would somehow always turn the conversation around to a point where she could say: "Did you know my son David is a city councilman?" That's my mom.

Why do parents always end up embarrassing their children? It didn't take long, though, for memories of my own behavior to surface. Reflecting on how I once snuck a photo of my own daughter's name on her office door made me chuckle and realize I could give my mom a break.

I guess it's just a parent's rite of passage. It's something you don't realize until you become a parent, but you grow as much as your children do over the years.

Winning the city council election was not something I expected but it certainly made me happy. I was even happier when I found out that the margin of victory was nearly eighty percent. This was an extremely hands-on position and gave me a real bird's eye view of the issues facing Taylor Lake Village and on some basic human truths, the first of which is that everyone wants to discuss only the issues that are affecting them.

Now, my community service went way beyond my little subdivision. The city council met twice a month in the Pasadena volunteer firehouse on Kirby Road. I was sworn in by the mayor and went right to work; I considered the job an expansion of what I had been doing for the Homeowners Association.

The role came with a small stipend for attending meetings and special sessions, but I made it a point not to take any payments from the city. I donated the stipend each month to a local community association or nonprofit. I was the only council member who never accepted any payment. But I was trying to give something back, not get something.

We were so fortunate to live in such a beautiful place, and to have raised our children there. I knew in my heart of hearts that if Amina and I had stayed in New York, it's possible our lives would have taken a totally different turn and we probably wouldn't have attained the same level of economic success or had such a nice place to base our family. I was grateful to the town, and could certainly do my part by contributing my time.

The first several months amounted to learning on the job as I immersed myself in our community, its citizens and politics. Never fear—small-time politics is alive and well in America.

One of my proudest endeavors was to be part of the organization to build the new city hall. The mayor, Jim Cumming, had led Taylor Lake Village for many years. In my opinion, he was one of the best. He truly understood our city, and knew the place inside and out, with the help of good people like Alice Riley, the city secretary and administrator.

Jim was a local authority, not just because he was mayor but also because he knew the city's history well. No one questioned him. At the time, building a new city hall wasn't a priority for some of the city council members but Jim knew we needed it. I was dedicated to the task as well and did everything I could to move it along.

When the item was finally placed on the agenda for a vote, it was exciting, and when it passed, it was a thrill. This was a big, important project, a tangible example of the good that local politicians can do.

On May 1, 1994, the City Hall was completed. A bronze plaque was placed on the building, with Jim's name as mayor, along with all the names of the members of the city council, including mine.

There it was—not exactly the same as having your name on a Broadway marquee, but, somehow, more special and important. What an honor and, really, how cool is that? I wasn't even dead and my name was immortalized. It will remain long after I'm gone. If I had known my name would be engraved in bronze and put on a building, I would have worked harder and faster to make it happen!

Of course, my mother came to Texas to see it, and made me take a picture standing in front of this plaque. I guess it's the kind of thing that runs in the family.

I was an active city council member, and had sponsored several ordinances. As time went on the mayor appointed me to the local police commission and, eventually, I became the chairman of the Lakeview Police Department.

Taylor Lake Village and another nearby city called El Lago each had operated their own municipal police departments. But in 1986 they decided to merge the departments as a way to save money.

At first, I thought my becoming chairman of the police department was a conflict of interest, since I was still a Taylor Lake Village councilman, and would have to vote on the department budget I had developed as police chairman.

The mayor and city attorney assured me this wasn't a problem; I proudly served in both positions, and the department budget was approved. Once the administrative hurdles were over, I went on to become more familiar with the police department and forge a close bond with several officers, including Lieutenant Howard, Officer Jason Smith, Sergeant Steve King, and many others. To understand what these officers were dealing with, I would ride along with several of them during their shifts.

The first time I did a "ride along" was memorable. I was in the car with Police Chief Tom Savage. We were traveling on Kirby Road, a long, straight stretch—and off in the distance a car was approaching kind of fast. Tom's radar showed the car was going thirty-nine miles per hour in a thirty-five-mile-per-hour zone—no biggie. And as the car got closer, you could see the driver slow down as she realized a police car was there, watching her.

Then as it slowed down to the proper speed and went by, I realized I knew that driver: my daughter, Lahna. I'm sure my surprise was nothing compared to the shock she probably got when she looked into the police car and saw her father sitting in the passenger seat!

By then Tom was explaining how people usually slow down once they see the police car, and he seemed to want to maybe stop it and give a warning. I didn't mention that the driver we had just seen was my daughter. I told him it was okay if we moved along. Little did he know Lahna would be getting a stern warning about obeying the speed limit when I got home.

I rode along several other times, once with Sergeant Steve King. We were on Kirby Road again when a dark, ragged-looking car with black tinted windows turned onto the road. I can't remember why Steve stopped him, but when he got out of the police cruiser and approached the vehicle, he left the two-way radio on so I could listen to what was happening. My heart was in my stomach as I sat in the car, wondering who this could be. Luckily, it turned out to be a Houston undercover officer, which was quite a relief!

Taylor Lake Village and El Lago weren't exactly hotbeds of criminal activity, but there was another incident while I was police chairman in which the department wasn't so lucky.

I had gone out to dinner and didn't take my phone or pager with me. When I got home, I went up to my office to tackle some left over work. Before I sat down I turned on my police scanner, like I normally did, and settled in to take care of my work. But after a few minutes it became clear that the chatter on the police scanner wasn't the normal variety. I knew immediately what was happening.

I called the dispatcher, Wendy Perez, who had been with the Lakeview PD for more than fifteen years (and is still the best dispatcher in Houston, in my opinion). As I dialed I suddenly became aware of the sound of helicopters and sirens in the immediate area, and I knew we were in the middle of a major incident. Immediately, I drove to the area.

Earlier in the day, someone was fishing. The person was on private land and refused to leave. The police were called and an officer was dispatched, just a normal call. The officer explained to the fisherman that he had to leave the private property—a simple request. The man probably wouldn't even have gotten a summons.

But it didn't turn out to be so simple. The man started challenging the officer and things quickly escalated until the man somehow pushed the officer into the lake.

The officer called for backup, and when he approached the man again he saw what every officer probably dreads: The man had a weapon. This officer put the man in a bear hug and threw him to the ground so he could try to take the weapon away.

When the backup officer arrived, he could see the struggle that was going on, and he pulled his own gun; he knew all of their lives were in danger and he did what he was trained to do. He had to use deadly force to contain the situation.

When I arrived, the chief filled me in and, as I stood there, all I could think of was how terrible it was. I was thanking God that none of the officers were hurt. But nothing could shake the terrible weight that seemed to hang in the air as the fisherman's body lay on the ground covered in a white sheet. It was awful.

Though we were in a small community where things like this weren't supposed to happen, this incident showed that sometimes it's the people who pass through your community that you have to worry about.

I spent some time afterwards with both of the officers who were involved in the shooting, but especially the one who had fired his weapon. He will always hold a special place in my thoughts. An incident like that weighs heavily in the heart and mind of any officer, and moving past it took time.

I always felt these men and women who protect us should be honored, and at the end of the year the department would have a banquet and invite all the officers, their families, the local officials and others. We would recognize their achievement and the police department would award a $500-savings bond, and Bay Area Travel would award two round-trip tickets on Continental for anywhere in the U.S.

Serving as police commissioner and getting to know those brave men and women was an honor. Being elected to the city council gave me more opportunities than I ever realized it would, and while I ran for and won a second term, I wasn't sure how long I wanted to be an elected official.

There were many people encouraging me to run for mayor. I was flattered, but I never thought of myself serving in that role. By the time an election for my third term in the city council came around, my name was on the ballot but I did nothing to promote myself. At that time there were several newcomers who were interested and campaigned actively. Even so, it was a close race. I lost by less than forty votes.

I stayed on as police commissioner. It was a term-limited position, one that I served in for six years. I never had aspirations, though, for a greater or more substantial political career. But you don't have to be president of the United States to make an impact. I do feel that the more people who get involved in the process, the better. We need new, creative, innovative young people who want to make their community a better place.

Throughout my community service, Bay Area Travel was going strong. One year we planned a trip to Hawaii on Continental Airlines—which had been taken over by Texas International—and decided to take my mother along. It was a direct flight from Houston on an A300 Airbus. This was the biggest equipment being flown at the time, and since Amina and I were doing so well professionally, we decided to make the trip special for everyone and booked three first-class tickets.

My mother always liked to peek inside the cockpit. In those days before the September 11 terror attacks, there was a much different attitude about security and things were more relaxed. I just happened to ask who the pilot was and, to my surprise, it was my old friend Captain Buddy Howard.

Once Buddy heard I was on board he came and gave me a big bear hug. We got so busy reminiscing about the past that I nearly forgot to introduce Buddy to Amina and my mother.

What made that meeting so special was seeing Buddy there, and realizing how far he had come. Just fifteen years after Texas International hired him, here he was. There was no higher-paying job, no better position, than to be the captain on an A300 Airbus. He was so happy, and it was joyful to see Buddy living his dream. He had accomplished so much and paved the road for so many minorities to follow.

When I heard recently that my friend Buddy Howard had passed away, I was shocked and very sad. I'm so proud to have crossed paths with such a dedicated professional.

If we're lucky, we start out our lives in the embrace of a loving family, and we get to go to school and have the opportunity to work and find our own place in the world. But so much of life is about what we learn along the way, and the people we meet can have as much influence on us as anything our parents tell us, or all the things we learn in childhood.

Getting to know Buddy Howard and taking a chance to make sure he had an opportunity at a pilot's job was a special experience for me. I made the decision to stand up for what I knew was right, and I'm so glad that I did.

Over the years, Bay Area Travel solidified its place in the Houston business community. We had room in our corporate headquarters, so we began a second business focusing on training people to work in the industry. We always focused on hiring the best people, and this gave us the chance to emphasize that philosophy in a new venture that was actually quite profitable.

Eventually, though, technology caught up with reality, and it allowed us to consolidate our offices. After about fifteen years of operation downtown, it was much more economical for us to operate under one roof, and we did that in the Clear Lake office.

How do you measure success? It's different for everyone. For me, it means being financially independent, and climbing out on a limb to start the travel agency with Amina is what helped us achieve personal security. At some point, though, it was time to move on.

The business was such a huge part of our lives, but with the years passing us by, and our children making us grandparents, we were ready for the next phase of our lives.

So we started to wind down Bay Area Travel, selling off parts of the business and shutting down others. It was time for another new frontier.

When you're thinking about retiring, it all sounds good: lazy days, freedom, travel. The day-to-day reality is different.

How many times can I walk around the mall? I thought. *How many times can I go to Home Depot without getting bored?*

Deep down I knew I would never be able to just sit at home. I had been working since I was thirteen years old; I wasn't going to be satisfied if I had nothing to do.

I decided a part-time job was the way to occupy my time. After all, I wasn't interested in the big bucks and didn't want a management position. I had been there and done that, and didn't need to drive the train anymore. I was fine with just finding a seat in the caboose and riding along.

But what could I do? I knew there were several Houston companies with opportunities available. I didn't want to be drawn into full-time work. It wasn't what I was looking for.

Years before, we had purchased a condo in New York City. It was on East 38th Street in Manhattan, just a few blocks from the United Nations in a fifty-seven-story building known as The Corinthian. Our unit was on the seventeenth floor, with a beautiful view of the East River and the Empire State Building. It was a magnificent view and a fantastic location, and I thank Amina for that, since she was the one who was pushing for us to buy in that building. I'm so glad we stayed the course and I'm so glad she was persistent.

When we bought the condo, our business was at its height and doing great. We had also bought a condo on the beach in Texas, on Galveston Island. Including our home, we had three properties—not bad for a kid whose first apartment was just an efficiency.

We tried each year to spend time at each of our properties, with our house as the home base. I would spend two months in Manhattan annually. New York, in some ways, will always be home. Just walking down the sidewalks and listening to the talk on the streets, I was amazed at how many dialects and languages are spoken there.

Manhattan is the financial capital of the world, and so diverse and busy. Yet even that wears off a little after a while, and we started to visit less often. Our relatives and friends had no problem requesting use of the condo, and I started to feel like a booking agent, coordinating schedules for the condo's use.

But when retirement came, I spent much more time there.

I think one year I saw every parade there was: from the Macy's Thanksgiving Day parade, to the Columbus Day and Puerto Rican Day parades, not to mention the Gay Pride parade and almost everything in between.

Why not put my love of the city to use? I decided to become a New York City tour guide, the ones that stand on those hop-on, hop-off, open-air tourists buses.

I studied for a month. You had to know Manhattan and all four of the city's outer boroughs, along with the bus and subway system, the architecture of the buildings, and all of the city's numerous parks. I succeeded. I am a licensed private tour guide for New York City. I always make sure my license is updated because I don't think my brain could handle that crazy test again.

Life kept pulling us back to Texas, where our hearts were. After a while we noticed we weren't using the New York condo as much. It was fully furnished, and our family and friends got quite a treat using it, but one day our Realtor approached us with an idea. She tried to talk us into doing a corporate lease of the place, and after some convincing we agreed to let her test the waters.

It didn't take long. One day she found what she had been looking for: a Westchester County man who lived in the suburbs and worked in Manhattan as a corporate CEO. He wanted the condo for limited use. I don't think he used the place more than two weeks out of the month, and was willing to pay close to $5,000 a month, which amazed me. It was too good for us to pass up.

With that, my retirement shifted away from New York and back to Houston. So much for being a tour guide.

We went back to Houston full-time, and that's when someone made a suggestion: Why not work at the space center?

Well, why not? That is where I started my career, after all. By now the Manned Spacecraft Center had been renamed the Johnson Space Center, after President Lyndon B. Johnson, and it was a top attraction. It seems everyone who wants to visit Houston visits NASA—after all, Houston is considered Space City, USA. It's America's gateway to the universe.

I started working there as a team member, but that didn't last long. Instead, I began doing something different. I started working as a level-9 tour guide.

Even though I had once worked with some of the most sensitive and critical components of the space program, securing work as a level-9 guide wasn't easy. I had to fill out an extensive application and undergo a thorough background check by the U.S. Department of Homeland Security. It took six months for my security clearance to be approved.

A level-9 tour guide at Space Center Houston is the highest-level NASA tour that's offered. We take people behind the scenes to see the world of NASA up close. The tour is a four-and-a-half- to five-hour experience that brings the public into places like historic mission control and other high-security spots at the space center, and tells the stories behind America's space program.

The Space Center Houston is a 180,000-square-foot complex that has entertained and informed more than 11 million star-struck guests, who come from every corner of the globe. We share the untold stories of the Apollo program and our country's historic journey to the moon.

First, we take you to lunch in the cafeteria that the astronauts eat in, which is in building 3. Inevitably we run into astronauts or crewmembers, which makes it a fun experience. Then, of course, all three mission control rooms: the one that the space shuttle is controlled from, the International Space

Station control room, and, of course, my favorite, the historic mission control room, where all the moon missions were controlled from, along with various other locations.

But what people really like is historic mission control. The tours accommodate just twelve people, plus a driver, along with the tour guide. When we step into Apollo mission control, we talk about the history of the room and everything that took place there. This mission control is on the National Register of Historic Places, and the equipment that people see there is not a mockup. These are the actual monitors and computers used during the Apollo program.

Because of that, one of the first things I tell people is not to touch anything, not just because of the historic nature of mission control, but also because we don't want to accidentally launch anything!

Historic mission control was the room that oversaw ten Gemini, ten Apollo and twenty-one shuttle flights, and returned them safely. The whole Space Center has this authenticity. When you walk down the halls today, you can hear the hollowness in the floors because they're actually computer floors, and underneath the floorboards is all the cabling that it took to run these computers.

There's enough wiring in the space center to orbit the earth three times. But that, along with the IBM 360 Model 75 mainframes, which took up a quarter of an acre, is what it took to get us to the moon.

"A man must rise above the Earth, to the top of the atmosphere and beyond, for only thus will he fully understand the world in which he lives."

Socrates said that in 399 B.C., and how forward-thinking it was. If you think about the equipment NASA has in that room, it's staggering. In March 1965, this was the world's most advanced technology at the time, amounting to 400 kilo-bytes—"megabyte" and "gigabyte" weren't even words back then.

It was a great privilege to be able to talk to a group on the day NASA celebrated the fortieth anniversary of Apollo 11 in July 2009.

I kept my eye on my watch as I spoke to the group about the Apollo program and its various missions. But, at one point, I made sure to stop.

"It's 3:18 p.m. (CST)," I said, pointing toward the clock before looking back at my audience.

Then I patted the console next to me and aimed my finger at the speaker.

"Right through this speaker, this same time forty years ago—some of us remember it—this is history. You're here at the exact time marking the moment when the words: 'Houston, Tranquility Base here. The Eagle has landed.' These historic words were spoken forty years ago, right through that speaker."

I told the group how 600 million people, about one-fifth of the world's population, watched the moon landing on televi-sion or listened to it on July 20, 1969.

It's funny, but sometimes people ask me whether we actu-ally did land on the moon. But those are the people that think Elvis is still alive and the earth is flat. My answer to that is pretty simple. If we didn't actually go to the moon like we said we did, who do you think the first people to reveal that would be? The Russians, of course, because we were in such a heated race with them to win the contest of who could conquer space first.

Besides that—Walter Cronkite would never lie.

Apollo 7, which flew on October 11, 1968, wasn't actually flown from this mission control. The first manned mission was actually controlled just one floor below. But it was also critical because the crew, Walter Schirra, Donn Eisele and Walt Cunningham, who I visited with on one of my tours recently, were flying so soon after the January 27, 1967, Apollo 1 fire that claimed the lives of astronauts Gus Grissom, Ed White and Roger Chaffee.

Thankfully, the Apollo program recovered from this tragedy and went on to have many more missions, each with their own stories.

Space might be a sterile environment, but Wally Schirra, one of the original Mercury 7 astronauts, got a head cold in space and gave it to the other crewmembers, making for an uncomfortable flight.

The crew of Apollo 8, Frank Borman, James Lovell and William Anders, got a special visit before their December 21, 1968, launch: historic airplane pilot Charles Lindbergh.

This was the first flight to achieve enough speed to escape the gravitational field of earth and the first to be captured by and escape from the gravitational field of another celestial body—the moon.

When Lindbergh visited the crew, he recalled how he held a piece of string up to a globe to measure the distance from New York City to Paris. He used that to figure out how much fuel he would need for the trip. The amount was equal to just a tenth of what Apollo 8 Saturn V would use every second.

Every crew would have a patch symbolizing their upcoming mission. The triangular shape of the Apollo 8 insignia represented the shape of the Apollo command module. It showed a figure eight looping around the earth and the moon. The number eight also represented their mission number.

The initial design was developed by Jim Lovell, who sketched it while riding in the backseat of a T-38 on a trip from California to Houston.

Crew patches are made into larger plaques and held in mission control until the mission has been successfully completed. The flight director picks one of the mission controllers who have made a contribution over and above the call of duty during the mission, and that person is given a special task. During an official ceremony with the crew, that person would have the privilege of hanging the mission plaque in its permanent resting place. It is considered a huge honor.

Talk about astronauts getting sick in space—Frank Borman one-upped Wally Schirra, that's for sure. On Apollo 8 he got sick inside the cockpit, on the way to and from the moon, which wasn't exactly pleasant as gravity was playing its part and making the evidence of his illness float around the cockpit!

Unfortunately, astronauts Jim McDivitt, David Scott and Rusty Schweickart, who launched on March 3, 1969, were in a similar situation after Schweickart also got sick. Eventually NASA had to change some of its mission objectives because of this trend.

Apollo 9 marked a significant step forward for the program, since it was the mission that proved the lunar module was a flight-worthy vehicle. This mission showed that the LM, which orbited the earth, was worth its $41 million-investment. The astronauts also tested and qualified new space suits that would be worn later by Neil Armstrong and Buzz Aldrin when they touched down on the surface of the moon.

Schweickart's illness helped NASA realize how important it was to give the astronauts a two-day rest after they'd achieved orbit to give them time to adjust before doing any extra vehicular activity. After all, it doesn't make for a good result when you're trying to test a new space suit and you may vomit inside of it. In fact, in the harsh environment of space, that could become life-threatening. Rusty Schweickart was the first lunar module pilot to test the LM in earth orbit. He was a nice guy; a few years later, we would do a joint tour together.

Apollo 10 marked the first time an all-veteran crew was flying. This was critical because the flight was serving as a full dress rehearsal for Apollo 11 and everyone was on watch for any snags or problems. They had to prove, finally, that the lunar module would do everything that it was designed to do, so that Apollo 11 would have clear sailing. If there had been an anomaly that couldn't be rectified, Apollo 11 probably wouldn't have made history.

As it turned out, that worst-case scenario nearly came to be when the crew of Apollo 10 ran into a problem. Their mission was to go to the moon, hover over the Sea of Tranquility and take pictures for Apollo 11.

Before the mission, one of the oxidizer tanks that was slated to go on Apollo 10 was dropped by a technician and slightly damaged. It was repaired and the oxidizer tanks were put on the shelf.

Apollo 10 was issued new oxidizer tanks and was flying a very specific mission: 100 hours, forty-three minutes and twenty-three seconds. The crew was to travel to a certain point, get as close to the moon as possible, about 47,000 feet and eight miles above the surface, and go no further.

One would think that getting so close must have been tempting for the Apollo 10 astronauts—if it had been me, I would have probably turned around to the person next to me and said: "Well, we're here…that Neil guy can get the next one." But, really, they couldn't have landed. The craft was too heavy at that point, and would have run out of fuel before coming home.

As the crew jettisoned their descent stage, something went wrong. The spacecraft started to catapult violently, turning end over end until the astronauts lost all sense of which way was up and which way was down.

Just in the nick of time, the veteran astronauts managed to stabilize their craft. An in-depth study done later by NASA showed they had been just two seconds from crashing on the surface of the moon.

Of course, many people know the story of Apollo 11. In addition to exploring the moon and taking samples of the surface soil and the rocks, we also left artifacts on the moon, as well as other items that made the trip with the astronauts to and from earth.

Deke Slayton, one of America's original Mercury astronauts, had once been grounded from space because of a medical condition. Before Apollo 1 was set to launch, that crew intended to give Slayton a diamond astronaut pin, a special gesture to signify that he would always be one of them.

But Apollo 1 was a great tragedy, and the pin was never given. It was alleged that the widows of the Apollo 1 crew made sure Slayton got his pin—and he, in turn, passed it along to Neil Armstrong, who placed it at the Sea of Tranquility.

Various other items were left as well, more than 100 artifacts, actually. We left a patch from Apollo 1 and another from Apollo 11, an olive branch, and a silicone disk that had an inscription from seventy-three nations with a message that was written only half the thickness of a human hair.

Then, we left several medals representing our greatest competitor at the time: the Russians. Yuri Gagarin and Vladimir Komarov had both been space pioneers and were killed; leaving the medals was a gesture of respect from one astronaut to another.

Now, we all know when we leave the country, there is customs paperwork to be completed on the return.

The crew of Apollo 11 never thought they would have to

complete standard customs declaration forms on their return from the moon, but they did. Most travelers declare art and fine jewelry, etc. However, the Apollo 11 crew declared a bunch of moon rocks and some dust.

At the time, the Russians were trying to beat us to the moon. It was the only time, up until that point, that the Russians shared their flight plans with NASA, so as to avoid an unnecessary collision in space.

Three days before Apollo 11 was to launch, the Russians launched an unmanned spacecraft to the moon. The whole point was to take the wind out of America's sails with an unmanned operation that would land on the surface of the moon before Armstrong and Aldrin, take pictures, and scoop up some samples of the surface soil before heading home. But something went wrong and the spacecraft crashed the next day, on July 21, while Armstrong and Aldrin were still on the moon.

By the time NASA launched Apollo 12, on November 14, 1969, we were pretty confident in the outcome. But Pete Conrad, Alan Bean and Richard Gordon ran into trouble just seconds after launch, when lightning hit the Saturn V rocket just thirty-six seconds and 2,000 meters after liftoff.

Another fifteen seconds after that, lightning struck again, at 4,400 meters. Suddenly all the telemetry was nonexistent and they were operating mostly on battery power. This was a dead spacecraft, with three crewmembers headed to the moon.

As the FD was reviewing all the options, a person sitting at a console in mission control had an idea. John Aaron had remembered there might be a way around this problem, since he had remembered a simulation of the problem. He told the flight director but the FD wasn't sure and referred him to the cap com.

Named for the capsule communicator, the cap com is an

astronaut who stays in mission control and communicates to the astronauts in space. It's felt that because they know all the acronyms and can talk to each other in their own unique way, it's better to have an astronaut to relay information to reduce the possibility of miscommunications.

So Aaron shared his idea through the FD to the cap com, who relayed it to the crew.

When Conrad heard the idea, he said: "What the heck?"

Well, maybe he didn't use the word "heck."

But rookie Alan Bean knew what switch it was, and as soon as he flipped it, everything was back to normal.

I had a personal interest in that flight. In November 1969, I signed a document along with others at the Manned Spacecraft Center, and the crew took the signatures on negatives and brought them to the moon. When they returned to earth they reproduced the signatures and gave them to me.

What blows my mind about that is the fact that they took a negative. There are probably plenty of people who think that a negative is simply the opposite of positive!

Apollo 13 nearly became an enormous tragedy for all of NASA, but ended up going on the books as a successful failure.

It started as a relatively easygoing flight, a routine mission of fifty-five hours, fifty-three minutes and eighteen seconds that traveled 199,000 miles away from earth, when suddenly there was an explosion.

The oxidizer tank on the crew's CSM had exploded. Now, the question you have to ask is: Where did Apollo 13 get their oxidizer tanks?

They were the repaired ones that had been put on the shelf from Apollo 10. The crew had signed off on using these but no one realized that, in 1966, when they changed the voltage from twenty-eight to sixty-five volts, no one adjusted the thermostats inside those tanks. When a certain switch got flipped, bam! That was all it took.

There was no way that astronauts James Lovell and Fred Haise could land on the moon after that. Now, they were in a potentially deadly situation.

"Houston, we have a problem."

Jack Swigert said those words first, but then Lovell came on and said them again. What we now know as a catchphrase was actually a statement of intense urgency.

That's when Gene Krantz said what is by now another popular phrase: "Failure is not an option."

The crew shifted themselves into the lunar module, and John Aaron, in his mid-twenties, would play a major role in powering Apollo 13 up and down.

The crew made it back alive because of the intelligent, thoughtful work of the people in mission control and many others behind the scenes. It was mostly north of the eyebrows—not because of the high-tech equipment.

A grateful crew returned to earth alive and well. When they came back to the space center, they had a gift: a mirror, which is hung on a wall near mission control with a plaque that reads: "This mirror, flown on Aquarius LM 7 to the moon April 11-17, 1970, returned by a grateful Apollo 13 crew to reflect the image of the people in mission control who got us back—Jim Lovell, John Swigert and Fred Haise."

Now, think of the number 13: the mission was named Apollo 13, launched 13:13 Houston time. The accident happened on April 13, and thirteen minutes after the oxygen tank exploded, they started to vent. Also the crew was launched on April 11, 1970. If you add 4-1-1-7-0, what do you get? Thirteen.

Ironically, NASA released a schedule of planned missions

recently that included one for June 2019, which would have
been the seventh human moon landing and the first lunar landing of the Constellation Program. President Barack Obama
cancelled the Constellation Program shortly after that.

Apollo 14 flew on January 31, 1971, with Alan Shepard as
the commander, and his crew, Stu Roosa and Ed Mitchell. It
was finally a normal mission, with Shepard hitting a golf ball
from the moon and then having to answer to NASA when he
returned.

On Apollo 16, which launched April 16, 1972, John Young
was the commander along with Ken Mattingly and Charlie
Duke as crewmembers. It was another smooth mission. I was
fortunate to be invited to the Cape to view the launch.

When you actually see and feel the roar of a Saturn V rocket
up close and personal, words are hard to find to describe the
moment—not to mention the feeling if you were an astronaut
sitting on top of it.

Thankfully, sometimes the problems astronauts encountered were less life-threatening than what Apollo 13 dealt with.
During Apollo 17, Commander Gene Cernan had noticed that
the fender on the lunar vehicle, NASA's own little moon buggy,
had been damaged.

I was with Gene Cernan when he was telling the story, a
couple of years after the December 1972 mission. If he had
been on earth, he explained to a group of Rockwell VIPs I had
brought over to the space center for a tour in 1974, he would
have just gotten a roll of duct tape and patched up the fender.

But he was in space. And with his pressurized helmet and
bulky suit and gloves, not to mention the dust, there was no
way he could just tear the tape like he normally would—with
his teeth.

Watching him mimic the roll of duct tape hitting the outside of his helmet was funny, and I still don't quite know how he managed to get the tape going so he could fix the fender, but he did. He's the one NASA astronaut that I always felt could do the near-impossible.

It's that kind of history, the big, important stuff along with the smaller, more human stories, that makes being a part of NASA so special. Commander Cernan knew it took the entire village to get him to the moon, which included all 400,000 Apollo workers. That's why his first step off the ladder was dedicated to the people who made it possible.

Working in the space center again is special to me. When I left Grumman all those years before, I didn't leave NASA, but, gradually, my life took a different path. Walking the halls of these buildings again, being around the astronauts and the engineers, sharing the stories of the Apollo program and the various missions—all of it was like coming home, in a way. I'm one of only a handful from that historic time who still work at NASA, giving me a certain notoriety, which I welcome.

I also conduct special protocol tours for NASA or the space center. Sometimes they're for the astronauts themselves and their families, or politicians with security details, or athletes. People always ask me about funny or unusual things that happen on one of these tours. I recently led a VIP group through the space center that included a man who identified himself as a federal agent.

Usually, if people who are carrying weapons come into the space center, security lets me know. But this time they didn't. The man alerted me when we were about to go through the metal detectors, and I let security handle it, according to NASA's procedures.

Everything was okay and we went along with the tour until we were in historic Mission Control. I was talking to the group about the Apollo landing and sharing all the memories when I heard a loud "thump" and nearly jumped out of my skin.

I turned around to see that the federal agent had fallen flat on his face.

Of course, he was on his feet immediately, and wasn't hurt, except his pride. He also asked not to do a report because it was not a big deal, and I agreed. But later, as I was talking about the moment with my friend Joe Cucci, he said something about people who carry concealed weapons, and it made me think, *What if this guy's weapon went off? He could have shot himself, or someone else, or even me.*

Can you imagine bullets flying around historic mission control? It wasn't something I wanted to think about.

I'm grateful that my blood brother and I still keep in touch on a regular basis.

Joe had a little more guidance than I did back in high school, and ended up going to a small college in the Midwest. I would end up doing it the hard way by going to college at night, after working all day. This is a good plan for how *not* to do it, and I work at encouraging school-age kids to start early, because you only want to load this hay once, and not on the back end.

I went on to live my life—working at NASA in the space program, in corporate America, launching my own business and raising a family. Joe went on to become a special agent for the Federal Bureau of Investigation in Chicago, where he had a thirty-one-year career.

I'm sure if a poll had been taken while we were both still in high school, we would have been voted as the two most likely *not* to succeed—with me nudging Joe out and taking the lead.

But things didn't turn out that way. I'm sure if we had stayed in Oyster Bay, Joe would have been driving a truck and I would have still been working at Nobman's. Now, there's nothing wrong with that. Yet we were able to find ways to achieve something more, and take advantage of opportunities that allowed us to rise above the meager expectations some people probably had for us.

Some of that is because we chose to leave New York. It's a great place, and many people discover their own paths to success in New York City and suburbs like Oyster Bay, where Joe and I grew up. Still, like anyplace, the area has its own unique limitations, and there is a big, open country out there with advantages for the taking. We happened to find our paths out there.

Occasionally, just Joe and I go back to visit the old neighborhood. About once a year we find ourselves in a New York state of mind, and return to the streets where we used to run around as children. We usually just walk the town up and down, remembering the past and talking about how things could have turned out differently. Sometimes we hook up with our old buddies, Billy Kruse, Jimmy Whitmarsh, Bill Schneider and Bill Jones. Bill was a popular name back then.

Like Joe, many of these guys ended up working in law enforcement. I always knew they would be involved with the law; I just wasn't sure what side.

It's amazing to walk around a place that shaped you so much and realize how far you've come, how small that place actually is, how much more there is in life. Joe always keeps me grounded.

Recently, Amina and I decided to downsize. Our children, Kumar and Lahna, were long since grown and out in homes of their own, but our dream home, built with all the balconies and porches, was now too big. It was approximately 6,600 square feet, had a pool, a built-in, 300-gallon fish tank, and 125 feet of bulkhead on the lake.

It was beautiful and we loved it, but it was becoming impractical. It was just too big for only the two of us. Sometimes, she would go up one staircase, I would go down another, and I never knew she was in the house until several hours later.

One day I was looking out to the backyard and started to count the maintenance hours and the costs. Amina and I realized a change was in order. We looked around and found a condo that was about to be built. It would be 3,200 square feet, and we would now be living vertically on three floors, with our own internal elevator. How cool is that?

The property is located in front of a ten-acre lake, which we share with other nearby town homes. Ironically, Jim West built this lake in the 1920s for his wife.

The West mansion is also located directly next door, and is now a national historic landmark. He donated more than 1,028 acres and that is the property NASA JSC sits on today. I bet he never expected to see me as his neighbor, or fishing out on his lake.

So we listed our home for sale and began building our new condominium. I've always had good karma when it came to real estate and I knew our house would sell, and it did.

We needed a place to stay while the condo was completed, so Realtor Janice Owens found us a location we could use for ninety days. Since it was such a short stay, we decided to wing it: no cable, no high-speed Internet, a television with aluminum foil on the antenna. Our old house had four bath-rooms just on the first floor, and this place only had one.

I was complaining to Joe about the lack of restrooms one day when he stopped me.

"Don't you remember the house you grew up in? You had eight people in your family, six kids and your parents—and only one bathroom. So shut up and stop whining," he said.

His point was valid and we both had a hearty laugh. I had to admit he was right. Only someone who knew me from the get-go could make that point.

■ ■ ■

From the people I've met to the work I've done, to the finan-cial success I've achieved—the most fulfilling thing is how that freedom has allowed me to give back, in so many ways.

Whenever time and money allowed, I made a point over the years to do something special with my mom. Even as a grown man with my own kids, I would check in with her every chance I got. After all, my "probation" for lifting those marbles never really had an expiration date.

But going on trips with my mom was something special. One year, we took her on a cruise to Bermuda, and another time, I took my mom to Opryland in Tennessee.

Now, don't ask me why my mother, born and raised in New York, always wanted to go to Opryland, but she did. I don't think I even heard country music myself until I came to Texas. But it was a surprise for Mother's Day, so it was a great treat for both her and me.

I was always careful about how and what I would do for my mom, since I didn't want any of my siblings to get upset. Every one of my siblings did their own thing with Mom, in their own way. I always asked her to not talk about any gifts I would send. But I felt strongly that I wanted to make sure to spend time with her, do things with her, while she still could. Life is precious, and I never wanted to ask myself: "Did I do everything that I could for my mother?"

We did a lot of different things. Sometimes, she would come to Texas. My mother loved Amina, and would always call her "my girl."

I don't think we can ever really do enough to thank our parents; at least we try as much as we can. I wanted so much to pay my mother back—to, in a way, help her experience all those things she missed out on.

One of the last special times I got to spend with my mother was in New York. It was early summer and she had never seen our condo in Manhattan.

Like a lot of New Yorkers, my mother had never gone to see a play on Broadway, either. So we went, just the two of us, and spent about four days in the city. She wanted me to cook one night, so I did, and we talked so much about so many things. We went to see *Phantom of the Opera*, and I wanted her to feel good so I hired a town car to pick us up at 38th Street and First Avenue, so we could make our way over to Broadway and 7th Avenue in style.

I remember that the musical was so emotional; during it, I looked over at my mother, who was sitting there in tranquility at her first Broadway show. I will always cherish how much she was enjoying it. Now, whenever I hear one of those songs, I tear up and will always remember that special moment and time with her.

When I got the phone call on July 10, 2002, that my mother had unexpectedly passed away, it rocked my world. I felt the loss acutely and miss her terribly, to this day. That's not the way

I wanted my "probation" to end, and it sent my world crashing down around me.

I was devastated, and now I knew how my own mother must have felt all those years before, when I was just a child sitting on the steps in the house, listening to her grieve for her own parent.

It was one of the most difficult times in my life. At that time I rededicated myself to making sure whatever I did in the future would make my mother proud.

As the years have gone by, I have been quietly amazed and so happy to see how my family turned out. Amina and I started our married life alone, no big ceremony, no crowded church.

I had always wanted Amina to get her church wedding, and we did so at Lakewood Church on our twentieth anniversary, attending a vow renewal ceremony performed by Pastor Ostein, Sr., with several other couples.

My family is a great source of pride, but I must say it gets a little confusing for me to explain my ethnic makeup to people—with my mother's Irish descent and my African-American father, along with the bit of Shinnecock Indian blood.

Well, my children have it even tougher. Amina's father, Sam, was from India and her mother, Mabel, is from the West Indies in Bermuda. So my kids, in a way, come from different ends of the world.

Kumar and Lahna are each married to Anglos, so Kumar's children, Xavier and Anika, who have Swiss-German heritage from their mother, Brooke, have a lot in their background.

And Lahna, now pregnant with her first baby, will have her work cut out for her in explaining it all to her child!

I often keep busy with my family, but am thrilled to have the time to give to the space center.

Bill Forster is a friend of the level-9 tour guides at Space

Center Houston. He works in mission control, as a lead ground controller and ascent-and-entry flight controller, and was on the console for thirty-six shuttle flights, including the Columbia disaster in 2003.

Bill helped enhance my training by giving me a chance to sit with him on the console at space shuttle mission control.

That day's simulation was a recreation of the December 09, 2006, flight of the STS-116 crew to the International Space Station. That crew helped to construct an outpost of the International Space Station and performed four spacewalks, which is no easy feat. Those astronauts also brought supplies and equipment to the space station, more than two tons' worth, and brought the equivalent amount back with them to earth.

The simulations are key training exercises for astronauts and the people working the missions. There is a special group of individuals who have to be certified and are experts in their field, and their job is to act as devil's advocates during training.

The work these people do is so important; they devise scenarios of all kinds, including system failures that the flight controllers and crew must be able to react to in a safe and timely manner. The simulation supervisor oversees this training, and there are around 6,800 possible malfunctions that can be activated during the exercises, making for a bad day for controllers and crew.

The simulation run that I sat in on was at night, so immediately after my level-9 tour I went to join Bill on the console at mission control.

I put on a headset, and settled in front of the console to see all the computer images flashing before me and hear several conversations being conducted at once. It was a hectic flurry of activity; everyone was in a rush but they were all confident in their approach.

Everyone responds to the flight director, who in turns relays the information to the cap com. After several hours of monitoring these activities, I realized, and freely admit, that this is a young person's game. Even with my new Lasik eyeballs, I still had to continuously lean forward to see the screen and the computer readouts.

After trying to follow all that activity through the end of the simulation, I was drained, and all I did was watch. These training exercises are critically important, and the people who run them are dedicated and talented individuals doing a great service for our country.

I've had a lot of experiences at the Johnson Space Center, and been around many intimate aspects of NASA's space program over the years. Whether as a technician working inside the actual lunar module, or as a tour guide sharing the special history of America's space exploration, I've seen and done an awful lot.

But on April 15, 2008, I got the ride of my life when I went with three other fellow tour guides to building 5. That's where NASA's $100 million-space shuttle motion-based simulator is kept.

Two of our coworkers, Jerry Hook and Irwin Stewart, couldn't make it that day.

So I went, at 8:54 a.m., with my other coworkers: Terry Hartman, a retired U.S. Air Force pilot; Georgene Harris, who had earned her mechanical engineering degree back when women weren't so active in that discipline; and Brenda Boykin, senior level-9 tour guide and sister-in-law of Dr. Bernard Harris, the first African-American to do an EVA space walk.

There we were—an Air Force pilot, a mechanical engineer,

a woman whose close relative had made NASA history…and me.

I have flown a small plane, now and then. And I had spent several hundred hours inside the lunar module cockpit, though it wasn't moving at the time and never left the ground while I was in it. So I was the lightweight of the group.

I wouldn't have passed up this opportunity for anything. Not everyone gets to walk in and use the same equipment that our astronauts do, or know what it feels like to land a space shuttle.

When we arrived, the simulation launch team went over what we would be doing, and then we were strapped in and given our headsets.

Here I was, laying flat on my back, looking straight up, snug in the five-point harness that I had been buckled into. It was so tightly strapped, I felt like I was about to be given a lethal dose of juice, just like they do on death row.

Then the cabin lights dimmed and my heart was throbbing. A voice started coming over the headset, and I could hear the team that was conducting this simulation explain that they were getting ready to start the countdown now.

10…9…8…7…

What am I doing? I thought. *This is crazy. I'm no astronaut! I'm no pilot!*

…4…3…2…1…launch!

Then there was no time to think. We had "launched." At T-minus 6.6 seconds, you could feel the "space shuttle's" main engine come to life, igniting with about seventeen percent of the initial thrust it would need.

Actually, each one of the engines fires, one at a time, 0.12 seconds apart. Then at T-minus zero, the bad boys start to light up as the solid rocket boosters kick in, delivering the rest of the eighty-three percent of the thrust our "shuttle" would need.

Now, we were rocking and rolling.

I never left earth, but I was being tossed around like a rag doll in there. Once the SRBs ignite, the shuttle is committed to launch. There's no going back.

At T-plus three seconds, the vehicle has reached full power, the special explosive bolts that I remember so well from my days working for Grumman go off, and the shuttle separates from the launch pad.

It was just a few blinks of an eye, but a handful of seconds never felt more like a lifetime. I was sitting in the commander's seat, on the left side of the cockpit. The other seat was occupied by another NASA simulator training pilot, who was really there just as a security blanket.

Our "launch" was so loud, and anything but silky smooth. There was so much movement and we really got tossed about. When the SRBs jettisoned just two minutes into the flight, which is their entire lifespan anyway, you could really feel it.

The solid rocket boosters fall away when the shuttle reaches an altitude of about twenty-eight miles; each SRB actually has eight small rockets that burn for just a second or so, but that helps the rocket booster fall away from the shuttle, and parachutes help slow the SRB's descent to the ocean. NASA's pretty conscious about recycling, and the SRBs are made to float for easy retrieval. They are typically reused on other flights. So if you plan any fishing trips on the day of a launch, you may want to reschedule.

I knew for sure that I was in a simulator. That I was in a machine that was still on the ground. But the effects of the simulator were very convincing. Inside that cockpit, I could really feel the SRBs falling away, could feel the real effects of a launch.

Now we were continuing on the thrust of the three main engines for about six and a half minutes, until our fuel was gone. Then the external fuel tank would also be jettisoned (though not recycled like an SRB).

Through all the noise and the shaking and different things going on, the programmed orbital maneuvering of the craft turns you upright and, finally, there is this amazing sense of serenity.

The rattling and screaming of the rocket boosters ends, and a quiet peace takes its place. You almost truly feel weightless, as if you could float freely around the cockpit if you wanted to.

I looked outside the window, but the illusion was complete. No one would ever convince me that I was not in earth's orbit. It was a beautiful time. I felt so light, and so lighthearted.

But space travel, simulated or otherwise, is not about peace. Reality quickly set in as the little birdie in my headset started chirping. The voice said: "You're in orbit now. We will perform the reentry."

They don't fool around: "launch," "orbit" and "reentry" are really all you get.

Now I would get a chance to shine. Reentry is no joking matter. We were to bring the orbiter down to approximately 50,000 feet, and then I had to give the command to remove the autopilot.

Then, it was all on me.

I would have to wrestle this flying brick to the ground, and hopefully make a safe landing. I was taking this quite seriously, as there were definitely bragging rights involved. How would it look, after all, if I crashed?

Usually, once you reach an altitude of fifty miles, the orbiter will hit some atmospheric drag. This is used to help the shuttle brake, and this descent, this crashing through the atmosphere, creates enormous heat.

The temperature usually peaks about forty-two minutes into the descent, when the speed of the shuttle reaches 15,000 miles per hour.

Believe me, you are boogying.

Normally at this point, because of the way the orbiter is ionized in the atmosphere, an astronaut would lose communications for about twelve minutes or so, but they skipped this part in the simulation and I'm sort of glad they did.

That's when they really separate the men from the boys, and I don't know what I would have done without any way to contact the outside world. The astronauts are so well-trained they can handle any situation that comes their way.

Finally I began my glide slope, which is when the shuttle begins to slant as it aims to make a landing. For the orbiter, this is seven times steeper than a commercial airplane approaching its landing. I felt as if I were leaning so far forward during this approach that if I weren't strapped in, I would have fallen out of the seat.

Now, NASA has what is called a Microwave Scanning Beam Landing System, which gives an autoland capability that can electronically find the orbiter and help it glide to a completely hands-off landing. So far, shuttle mission commanders have taken control of the orbiters during subsonic flight, usually about twenty-two miles from touchdown.

No one reminded me of that. Halfway through my reentry approach, my heart was beating double time and if I had remembered, I would have opted for it.

Usually, the sequence for the shuttle approach and landing is very specific. At about eighty-six seconds to touchdown, or

7.5 miles until the runway, the shuttle is zooming at 424 miles per hour and at an altitude of 1,000 feet. That's when you initiate the pre-flare—twenty-degree glide slope.

At thirty-two seconds to touchdown, just two miles from the runway, the shuttle is still speeding along at 351 miles per hour and is 1,747 feet high in the air.

A few breaths later, with only seventeeen seconds between you and the runway, the shuttle has only 3,540 feet to go and is still traveling at 308 miles per hour. It was just 134 feet up by now, and I was just focused on one thing: my heads up display.

Just like in the movies, a heads up display (HUD) is transparent and lets you look at it while keeping your head up and focused on where you have to go.

There were so many things going on, and when you're coming in for a final approach, the HUD makes for a much more accurate landing, especially for a spacecraft.

Every cell in my body was focused on that display. I blocked out everything else as my senses seemed to absorb the information from the screen. I never realized I could be so transformed, so consumed with a single task, but all I could think about was a safe landing.

You put the wheels down with fourteen seconds to go, and I signaled to the pilot position in the right seat to deploy the landing gear. Just 1,000 feet to go until the runway, and 267 miles per hour of momentum and speed carried me along.

On a normal shuttle landing, with a trained crew, the touchdown would happen at about 2,261 feet from the end of the runway and going 215 miles per hour. When I "landed," my groundspeed was approximately 172 miles per hour (the slower the better for me!) but I was still moving unbelievably fast—faster than anything I had ever been in control of before.

But it wasn't over yet. Just because the shuttle had touched down didn't mean I was finished. You're still in the thick of things because the orbiter drag shoots have to be activated, and the two main landing gears are in contact with the runway.

I didn't realize that every possible bit of information from my "flight"—from altitude above the runway to the tire pressure—was being measured and tracked. I thought if perhaps something had gone wrong, that it would just be something between myself and a couple of simulator guys. Boy, was I wrong.

My two main landing gears came into contact with the runway at around 160 nautical miles per hour. The tires under the nose gear made contact and the parachutes were then deployed and fully inflated to help finally slow down this massive machine, until the shuttle hit about thirty nautical miles per hour and was allowed to roll to a stop.

That's when I could finally exhale.

I realized it was over and I can say unequivocally that I know that I have not done, or will ever do for the rest of my life, anything more rewarding than flying and landing the space shuttle—or a simulation of one.

After all the years I have spent working at NASA, from 1968 during the Apollo program until now, I do feel I have come *full circle.* I realize only a select group of people get to do what I have done. I thank God I had the chance and was chosen for the experience.

As I walked out of the shuttle simulator, feeling about two inches taller, I went up to the training supervisor, who went over my stats with me. He congratulated me on a great flight and landing, which made me feel like I was on top of the moon.

Then I noticed that my colleague, Terry Hartman, was on his final approach in the simulator. So I took matters into my own hands.

I casually mentioned to the training supervisor that maybe Terry shouldn't have been on the same flight program as me. After all, he was a retired Air Force military combat pilot! He had more than 3,200 flight hours in a variety of aircraft. Terry had flown practically everything that hadn't been nailed down, including helicopters.

I kept the heat on with the supervisor, trying to fill his head with all kinds of bad thoughts.

"You know, Terry is one of those hotshot flyboys," I said. "You know these guys, you know how they are. We have to mix it up for him, challenge him a little bit."

I must have touched a nerve because the supervisor had no problem accommodating some of my suggestions. He immediately played along, sending the computer commands that changed Terry's "day" landing into a "night" landing and throwing other curveballs. Terry managed it all, of course, and when he got out he took one look over at me standing by the console and realized what had happened. We ended the day with a great big belly laugh.

My work as a tour guide has brought me into contact with all kinds of different people, including those at NASA that I admire the most. Charlie Bolden was one of those people. Charlie was a retired U.S. Marine Corps major general, a former Marine aviator and an astronaut, and, in May 2009, President Barack Obama appointed him as the first African-American administrator of NASA.

Before he was nominated, Charlie and his wife, Jackie, invited Amina and me along to an event, and while we were all chatting, Jackie asked me what I was up to. I started to explain how I had retired, and was now working part-time at NASA. Then she stopped me.

"No, David, you're like Charlie. You never say 'retired.' It's called 'retransitioning,'" she said, and laughed.

That might be true, but we're certainly at two different ends of the spectrum. I'm confident my friend Charlie is the right man to lead NASA into the future.

I'll never forget one special tour that ended at a dinner.

I have great respect for people like George Abbey. He held so many important positions at headquarters and at the Johnson Space Center and was so well-known that all you would have to do is say his first name and everyone knew whom you were talking about. He became the director of the space center in 1996.

One day I had completed a tour for a group of European high school students. They were all scholars and belonged to a prestigious space school. They had competed for the chance to visit NASA and were all extremely intelligent.

After the day-long tour, we all met at a local restaurant, the Villa Capri. George came along, and, as a special treat, so did Yuri Gidzenko, one of the very first crewmembers on board the International Space Station.

These were prestigious, accomplished men. After dinner, the three of us were standing together and the kids started to line up, wanting to shake our hands.

Then one of the kids came off the line. He was a skinny, frail kid—about the size I was when I was his age. He came right to me, passing up George and Yuri.

He asked for my autograph.

The heat started to creep into my face and I'm sure everyone could see that I was blushing as I tried to explain to him how important George and Yuri were—the man who actually ran NASA JSC, and, of course, the Russian cosmonaut!

The boy politely looked over, surveying both George and Yuri, but then he just looked straight at me. His eyes locked onto mine, and his sincerity and intensity sent chills through me as he said bluntly: "I'm probably never going to be an astronaut, or run NASA as a director. But I want to work in the space program, like you did and so many others."

Thankfully, it wasn't long before the group was asking for autographs from all three of us. At one point I leaned over to George and whispered: "Seems like these young people finally got the pecking order right." He chuckled, and I always enjoyed whenever I made George laugh, since it was a rarity.

That young man touched my heart in a very special way.

I was just a technician, but because of all that I'd achieved in my life and because I shared my story, this boy felt like he could be a part of NASA someday. He wanted a seat at the table, and seeing me made him feel like one day he could find that spot.

I was just a man. I have no special education. I come from a humble background, from a family that struggled to stay afloat and parents who couldn't really give me more than their love and their sense of responsibility, of right and wrong.

Yet my life is a success, in many definitions of the word. I have a strong, loving marriage, beautiful children and grandchildren, found a place in corporate America and also made my mark as an entrepreneur.

But this kid didn't know much about all that. He just saw

someone who found his own path amid a stellar group of high-powered specialists, astronauts and engineers at NASA.

He saw that if I could do it, so could anyone—so could he.

That would become one of my finest achievements, which I know would make my mother, Elnora Fredana Cunningham, very proud. Mom, I loved you then, I love you now, and I will always love you for being there for me.